Conference Proceedings of the Society for Experimental Mechanics Series

Series Editor

Kristin B. Zimmerman, Ph.D.
Society for Experimental Mechanics, Inc.,
Bethel, CT, USA

The Conference Proceedings of the Society for Experimental Mechanics Series presents early findings and case studies from a wide range of fundamental and applied work across the broad range of fields that comprise Experimental Mechanics. Series volumes follow the principle tracks or focus topics featured in each of the Society's two annual conferences: IMAC, A Conference and Exposition on Structural Dynamics, and the Society's Annual Conference & Exposition and will address critical areas of interest to researchers and design engineers working in all areas of Structural Dynamics, Solid Mechanics and Materials Research.

More information about this series at http://www.springer.com/series/8922

Ming-Tzer Lin • Cosme Furlong • Chi-Hung Hwang
Editors

Advancement of Optical Methods & Digital Image Correlation in Experimental Mechanics

Proceedings of the 2020 Annual Conference on Experimental and Applied Mechanics

Springer

Editors
Ming-Tzer Lin
National Chung Hsing University
Taichung, Taiwan

Chi-Hung Hwang
Taiwan Instrument Technology Institute/NARLabs
Hsinchu, Taiwan

Cosme Furlong
Mechanical Engineering Department
Worcester Polytechnic Institute
Worcester, MA, USA

ISSN 2191-5644 ISSN 2191-5652 (electronic)
Conference Proceedings of the Society for Experimental Mechanics Series
ISBN 978-3-030-59775-7 ISBN 978-3-030-59773-3 (eBook)
https://doi.org/10.1007/978-3-030-59773-3

This Springer imprint is published by the registered company Springer Nature Switzerland AG
The registered company address is: Gewerbestrasse 11, 6330 Cham, Switzerland

Preface

Advancement of Optical Methods & Digital Image Correlation in Experimental Mechanics represents one of seven volumes of technical papers to be presented at the SEM 2020 SEM Annual Conference & Exposition on Experimental and Applied Mechanics organized by the Society for Experimental Mechanics scheduled to be held in Orlando, FL, September 14–17, 2020. The complete proceedings also include the following volumes: *Dynamic Behavior of Materials; Challenges in Mechanics of Time-Dependent Material*; *Fracture, Fatigue, Failure and Damage Evolution*; *Mechanics of Biological Systems and Materials, Micro- and Nanomechanics & Research Applications*; *Mechanics of Composite, Hybrid & Multifunctional Materials*; and *Thermomechanics & Infrared Imaging, Inverse Problem Methodologies and Mechanics of Additive & Advanced Manufactured Materials.*

Each collection presents early findings from experimental and computational investigations on an important area within experimental mechanics, with optical methods and digital image correlation (DIC) being important areas.

With the advancement in imaging instrumentation, lighting resources, computational power, and data storage, optical methods have gained wide applications across the experimental mechanics society during the past decades. These methods have been applied for measurements over a wide range of spatial domain and temporal resolution. Optical methods have utilized a full-range of wavelengths from X-ray to visible light and infrared. They have been developed not only to make two-dimensional and three-dimensional deformation measurements on the surface, but also to make volumetric measurements throughout the interior of a material body.

The area of digital image correlation has been an integral track within the SEM Annual Conference spearheaded by Professor Michael Sutton from the University of South Carolina. The contributed chapters within this section of the volume span technical aspects of DIC.

The conference organizers thank the authors, presenters, and session chairs for their participation, support, and contribution to this very exciting area of experimental mechanics.

Taichung, Taiwan — Ming-Tzer Lin
Worcester, USA — Cosme Furlong
Hsinchu, Taiwan — C.-H. Hwang

Contents

1 **Diagnosis of Deformation Stages with Optical Interferometric Technique and Comprehensive Theory of Deformation and Fracture** 1
Sanichiro Yoshida and Tomohiro Sasaki

2 **Non-contact Measurement of Strains Using Two Orthogonal Sets of Twin "Blue" Lasers** 7
R. W. L. Fong and J. Patrick

3 **Experimental Observations on the Fracture of Metals** 19
C. A. Sciammarella, L. Lamberti, and F. M. Sciammarella

4 **A Digital Laser Speckle Technique for Generating Slope, Curvature, and Deflection Contours of Bent Plates** 43
Austin Giordano and Fu-Pen Chiang

5 **Holography and Holographic Interferometry via Photopolymer Film** 51
Austin Giordano, Lionel T. Keene, Ryan Norris, and Fu-Pen Chiang

6 **Evaluating Stresses from Measured Strains in Viscoelastic Body Using Numerical Laplace Transformation**. 55
S. Taguchi, K. Takeo, and S. Yoneyama

7 **Evaluation of the Influence of Water Absorptivity on the Properties of CFRP Cylinder Materials by SHPB Impact Compression Test** 61
J. Liu, K. Takeo, and S. Yoneyama

8 **Speckling and Testing with DIC at Microscales** 67
Kevin B. Connolly and W. Carter Ralph

9 **Perspective Compensation of 2D-DIC Measurements by Combination with Speckle Imaging** 73
Juuso Heikkinen and Gary S. Schajer

10 **Holographic Measurement of Semi-transparent Tympanic Membrane Shape Using Multiple Angle Illuminations** 79
H. Tang, P. Psota, J. J. Rosowski, J. T. Cheng, and C. Furlong

11 **Characterization of Interface Debonding Behavior Utilizing an Embedded Digital Image Correlation Scheme** 83
Tomislav Kosta and Jesus O. Mares Jr.

12 **Preliminary Characterization of a Plastic Piezoelectric Motor Stator Using High-Speed Digital Holographic Interferometry** 89
Z. Zhao, P. A. Carvalho, H. Tang, K. Pooladvand, K. Y. Gandomi, C. J. Nycz, C. Furlong, and G. S. Fischer

13 **DIC Measurement of Anisotropy for Plastically Deformed Thermoplastic** 95
Kenichi Sakaue and Sho Higuchi

Chapter 1
Diagnosis of Deformation Stages with Optical Interferometric Technique and Comprehensive Theory of Deformation and Fracture

Sanichiro Yoshida and Tomohiro Sasaki

Abstract A method to diagnose the stages of deformation nondestructively, quickly and as full-field information is discussed. In-plane sensitive Electronic Speckle-Pattern Interferometry is used in the subtraction mode to form the fringe pattern representing the differential displacement occurring during a short-time interval. The dark fringes in a fringe pattern exhibit the contours of the differential displacement field. The interferometer keeps forming such fringe patterns continuously. The change in the fringe pattern with the development of the deformation is interpreted based on a field theory of deformation and fracture. Based on a fundamental principle of physics, this theory describes all stages of deformation and fracture on the same theoretical basis. It does not need to use phenomenology or empirical formulation of the phenomenon. The transition from one stage to another, e.g., the elastic to plastic stage or the plastic to fracturing stage, of given deformation is diagnosed based on specific features of the fringe patterns and the field theoretical interpretation of the features. The transition from the elastic to plastic stage is characterized by the generation of shear instability that triggers the initiation of a large-scale rotation wave. The transition from the plastic to fracturing stage is characterized by the immobility of the rotation wave that causes the generation of material discontinuity.

Keywords Deformation theory · Electronic Speckle-Pattern Interferometry · Nondestructive testing · Field theory

1.1 Introduction

For nondestructive evaluations of solid objects, diagnosis of transition from one stage to another is important. Generally, the transition from the elastic to plastic regime is identified as the change from the linear to nonlinear behavior of the stress-strain characteristics. In field applications, it is unrealistic to measure stress-strain characteristics. While various techniques are available for deformation (strain) measurement [1–3], stress is hard to measure in the field. It is desirable to diagnose the current deformation status from the spatiotemporal characteristics of strain. A challenge for this approach is the lack of theory that describes all stages of deformation comprehensively. Prevailing practice is to use stage-specific theories: theories of elasticity [4], plasticity [5], and fracture mechanics [6] in the respective regimes. This is mainly because these theories are based on phenomenology.

The application of a recent field theory [7] of deformation and fracture can be a solution to the problem. Being based on the physical principle known as the local symmetry of theory [8], this theory has a mechanism to describe all stages of deformation on the same theoretical basis without relying on phenomenology. It formulates deformation as dynamics of mechanical waves that carry stress energy through the material. Fracture is characterized as the final stage of deformation where the mechanical wave is unable to carry the stress energy, and consequently, the stress becomes stationary at a certain location of the material. At this stage of deformation, the only mechanism for the material to establish energy balance becomes the generation of discontinuity, which causes the fracture. Previously various experimental studies [9–14] exhibit evidence that evolution of deformation can be characterized by different forms of the displacement wave as explained by

S. Yoshida (✉)
Department of Chemistry and Physics, Southeastern Louisiana University, Hammond, LA, USA
e-mail: syoshida@selu.edu

T. Sasaki
Department of Mechanical Engineering, Niigata University, Niigata-shi, Niigata, Japan
e-mail: tomodx@eng.niigata-u.ac.jp

M.-T. Lin et al. (eds.), *Advancement of Optical Methods & Digital Image Correlation in Experimental Mechanics*, Conference Proceedings of the Society for Experimental Mechanics Series, https://doi.org/10.1007/978-3-030-59773-3_1

the field theory. In this study, we numerically solve the wave equation derived by the field theory and compare the resultant behavior of the displacement field with experimental results.

1.2 Theoretical Background

Detailed description of the field theory can be found elsewhere [7]. In short, the theory derives a set of field equations that govern the elastoplastic dynamics of solid materials based on the physical principle known as local symmetry [8] and associated Lagrangian formalism. The wave equation can be put in the following form.

$$\frac{\partial^2 \vec{v}}{\partial t^2} + \sigma \left(\nabla \cdot \frac{\partial \vec{v}}{\partial t} \right) \frac{\partial \vec{v}}{\partial t} - \frac{G}{\rho} \nabla^2 \vec{v} = -\frac{G}{\rho} \nabla \left(\nabla \cdot \vec{v} \right) + \alpha \frac{\lambda + 2G}{\rho} \nabla \left(\nabla \cdot \vec{v} \right) \tag{1.1}$$

Here $\vec{v}$ is the differential displacement vector, σ is the material constant that represents the energy dissipative nature, G is the shear modulus, α is a parameter that indicates the degree of elasticity in the elastoplastic regime ($0 < \alpha < 1$), and λ is Lamé's first parameter.

When wave Equation (1.1) is expressed in the principal coordinate system, the wave solution in each regime can be classified as shown in Table 1.1. In the linear elastic regime, the differential displacement field exhibits the well-known compression wave characteristics. In the principal coordinate system, the distortion tensor does not have shear components. In the elastoplastic regime, the distortion tensor components start to take non-zero values. This excites transverse waves in the differential displacement field. The irreversibility due to plasticity is reflected in the wave dynamics as the fact that the longitudinal and transverse waves decay. In the pre-fracture regime, the wave is concentrated in a certain location of the specimen and travels as a solitary wave. When the solitary wave becomes stationary, the fracture is in the final stage, and material discontinuity is generated at the point where the solitary wave becomes stationary.

1.3 Results and Discussions

Figure 1.1 illustrates the experimental arrangement. A plate specimen is attached to a test machine for loadings. An in-plane sensitive dual-beam Electronic Speckle-Pattern Interferometer (ESPI) is configured to measure the differential displacement continuously as the test machine applies the load. At each time step, the current image is subtracted from the image taken in a previous time step. This generates the so-called subtraction fringe pattern as indicated in Fig. 1.1. The dark fringes in the fringe pattern represent the contour of the differential displacement occurring during the time interval between the two time steps involved in the image subtraction.

Figure 1.2 shows typical fringe patterns obtained by the ESPI setup where a tensile load at a constant pulling rate is applied to the specimen (aluminum alloy A5083, 2%Mg). Labels (a)–(c) indicate the location on the stress-strain curve at which each group of fringe patterns are formed. The dark fringes represent the contour of the differential displacement vector component parallel to the applied load.

The fringe patterns (a)–(c) can be characterized as follows. (a) It exhibits a uniformly distanced dark fringes running approximately perpendicular to the tensile axis. Since the dark fringes represent the differential displacement component parallel to the tensile axis, this pattern of the dark fringes indicate that the wave is longitudinal. At stage (b), the dark fringes become tilted to the tensile axis, indicating that the shear components become non-zero in the strain tensor. It also indicates that a transverse wave is being exited at this stage. The fringe pattern in stage (c) consists of curved dark fringes sandwiching parallel, slant linear fringes. These slant linear fringes can be interpreted as representing a solitary wave [7]. As the three fringe images indicate, the solitary wave travels along the specimen. The fact that the solitary wave is dynamic indicates that the specimen is not ready for the final fracture.

Figure 1.3 shows fringe patterns from the same type of experiment as Fig. 1.2. In this case, both the displacement vector components parallel and perpendicular to the tensile axis were measured with an ESPI setup similar to Fig. 1.1

Table 1.1 Forms of differential displacement wave

Regime	Linear elastic	Elastoplastic	Pre-fracturing
Wave	Longitudinal waves	Decaying longitudinal/transverse waves	Solitary waves

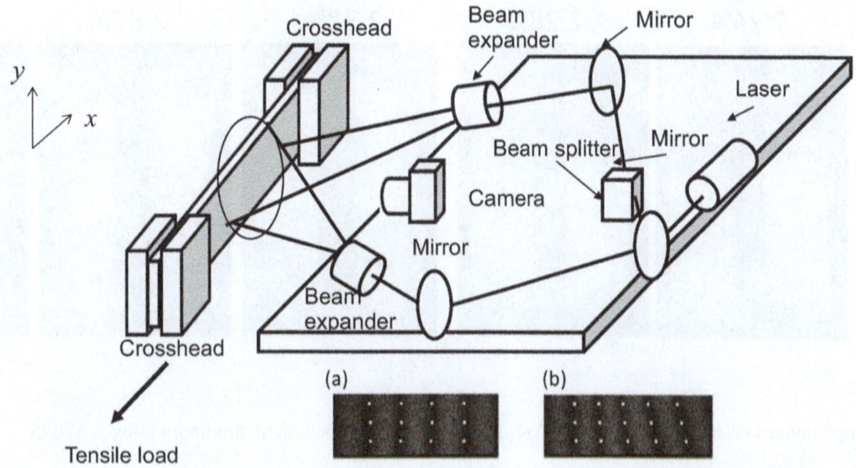

Fig. 1.1 Experimental setup. (**a**) and (**b**) are typical fringe patterns. The dark fringes are highlighted with white dashed lines

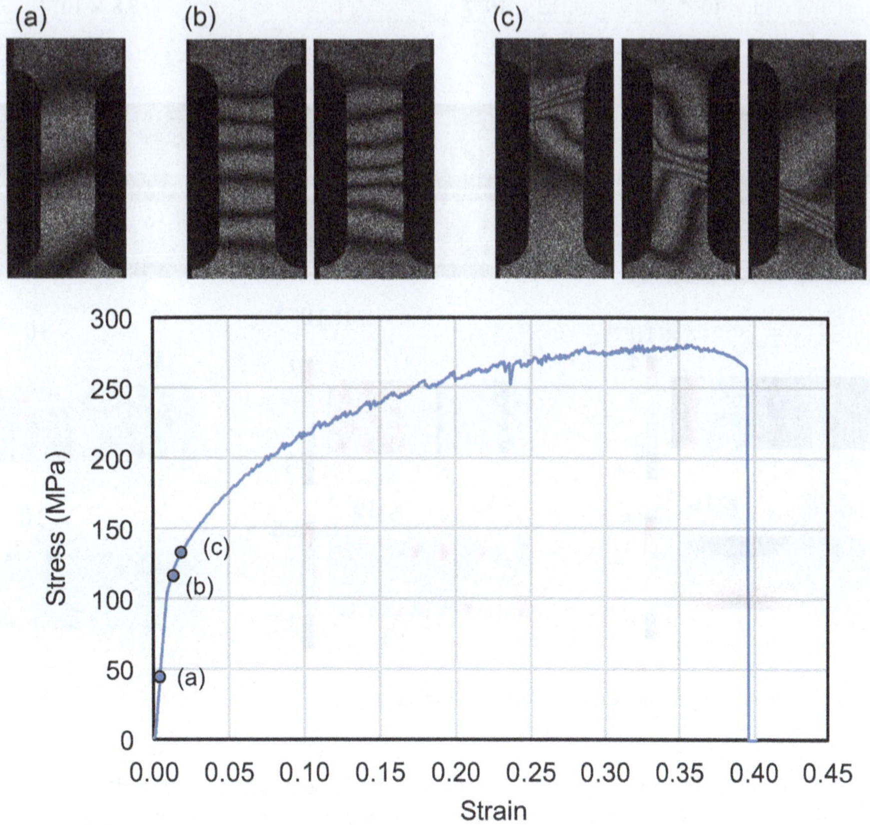

Fig. 1.2 Typical fringe patterns representing longitudinal differential displacement at different locations on stress-strain curve

where an interferometer sensitive to the in-plane displacement component perpendicular to the tensile axis is added to the configuration. For each pair of images, the left represents the differential displacement component parallel to the tensile axis (v_x) and the right represents the component perpendicular (v_y). The number above each pair represents the normal strain in the direction of tensile axis. The yield strain was 1.0%. It is seen that the pattern travels at a constant rate along the tensile axis.

An important question raised in Fig. 1.3 is if the patterns represent the wave's characteristics given as solutions to wave Equation (1.1). To examine this question, a finite element analysis was made. This numerical model simply solves wave Equation (1.1), without using any constitutive equation except for Poisson's ratio. Table 1.2 lists the parameters used for

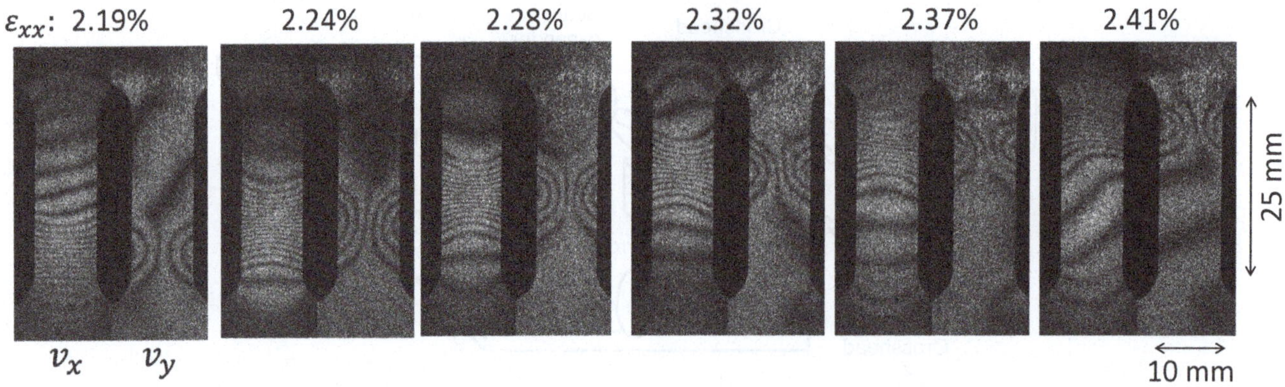

Fig. 1.3 Successive fringe images taken post yield points. Yield strain is 1.00%. Specimen: aluminum alloy AA7075

Table 1.2 Parameters used for numerical analysis

Pulling rate (mm/s)	σ (1/s)	G (N/m^2)	ρ (kg/m^3)	λ (N/m^2)	α
0.45	1×10^{-8}	2.2×10^{-4}	1	8.8×10^{-4}	0

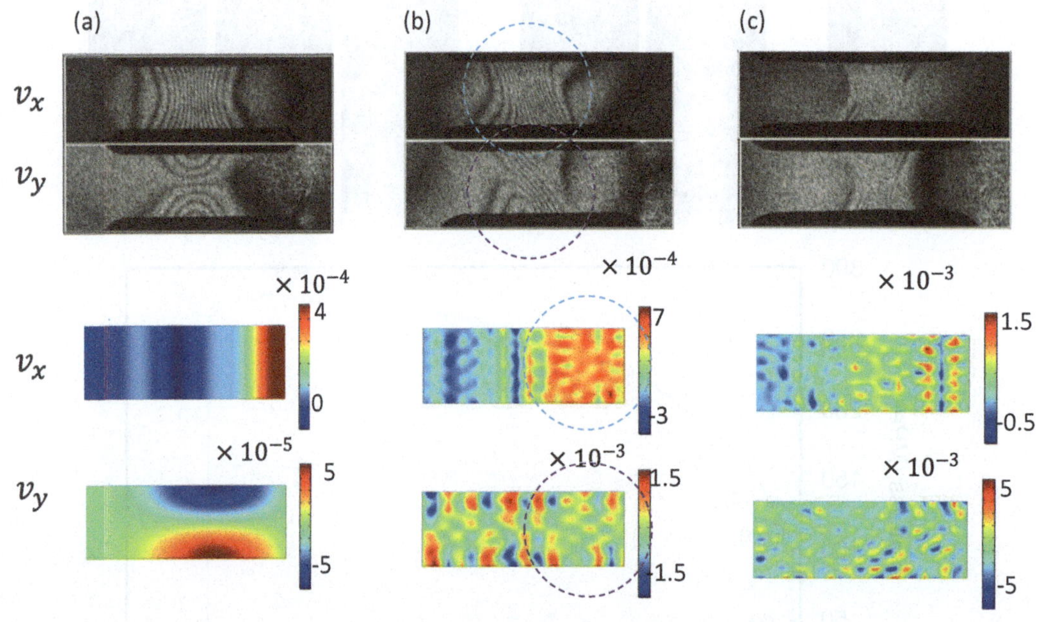

Fig. 1.4 Experimental fringe patterns (upper) and numerical analysis (lower)

the numerical analysis. Note that this condition was assimilated to the experiment through the wave velocity. The set of parameters shown in Table 1.2 provides the transverse wave velocity close to experimental values [11].

Figure 1.4 compares the experimental fringe images and the numerical patterns of v_x and v_y for three representative stages. Stage (a) is when the normal strain is approximately 2% (the same stage as Fig. 1.3). In this stage, the v_x and v_y patterns move at a constant speed. The numerical patterns clearly indicate the feature that v_x exhibits longitudinal wave characteristics where the contours are approximately perpendicular to the tensile axis and v_y exhibits semicircular pattern as shown. Careful examination reveals that the numerical v_x pattern is denser near the right end of the specimen. This seems to correspond to the concentrated fringes near the middle of the specimen observed in the experimental v_x pattern. This indicates that the numerical result shows strain concentration although its location is not the same as experiment. It is likely that the strain concentration occurs at a crest of the longitudinal wave in v_x.

The experimental image at stage (b) was taken when the normal strain is approximately 3%. At this stage, the v_x fringes are more concentrated and curved as compared with stage (a). The v_y fringes are not semi-circular. Instead, they are rather

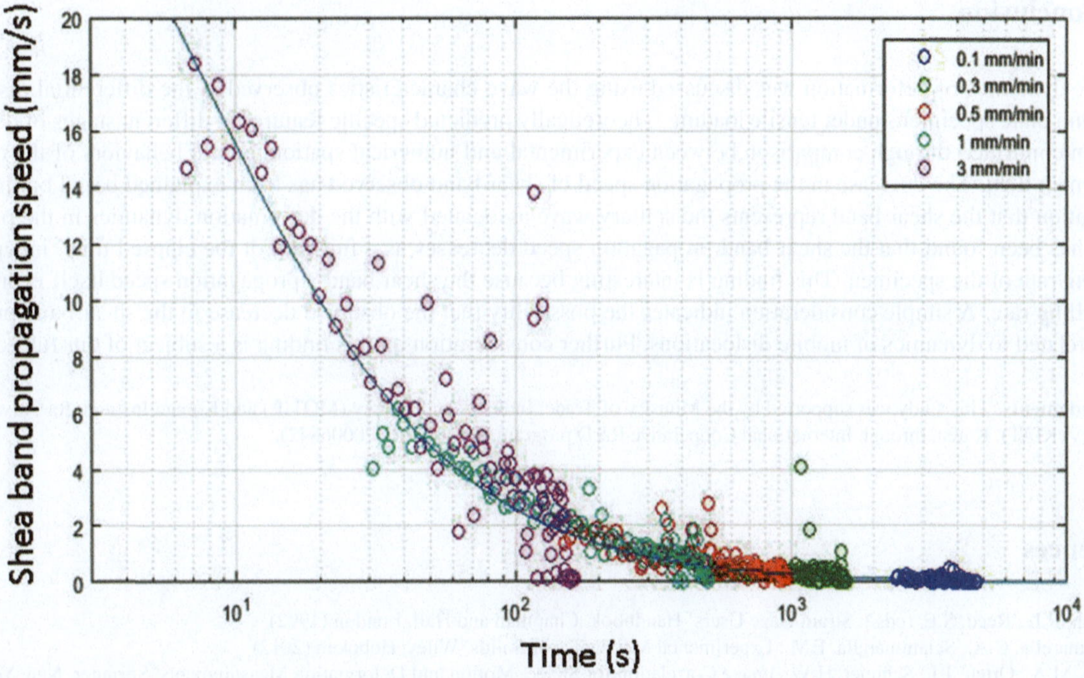

Fig. 1.5 Shear band propagation speed as a function of time for different pulling rates

slant and linear. The numerical v_x pattern indicates the more concentrated feature, and the numerical v_y pattern shows slanted feature. These features are consistent with the experimental patterns (see the areas enclosed by a circle).

In a later stage (c), when the normal strain is approximately 18% and the stress reaches the peak value on the stress-strain curve from where it decreases gradually, the experimental patterns of v_x and v_y are more concentrated and become similar to each other. A recent study [14] indicates that the similarity between the differential displacement component reveals that the pre-fracture condition is about to establish.

The solitary wave representing the pre-fracture stage appears in a fringe image as a shear band. A theoretical consideration [15] indicates that in the case of tensile loading, the solitary wave velocity is proportional to the pulling rate. It has also been found that when the solitary wave velocity becomes zero, the volume expansion rate of the unit volume becomes infinite and that causes material discontinuity. These indicate that the speed of shear band is fundamentally related to the fracture mechanism, and it is worthwhile considering more deeply. Figure 1.5 [16] plots the propagation speed of the shear band as a function of time for four different pulling rates. The horizontal axis represents the time elapsed from the beginning of the tensile loading. Two features are seen. First, the higher the pulling rate, the earlier the shear band starts to appear. It should be noted that the appearance of the first shear band (the onset time) is not necessarily determined by the total elongation. When the pulling rate is 3 mm/min, the first shear band appears approximately in 6.6 s. Inversely proportional to the pulling rate, the same total elongation occurs around 19.8 s ($6.6 \times 3 = 19.8$), 39.6 s, 66 s, and 198 s for the pulling rates of 1, 0.5, 0.3, and 0.1 mm/min, respectively. However, Fig. 1.5 indicates that the onset times for these pulling rates are 30 s, 150 s, 400 s, and 2400 s. Apparently, the actual onset times are greater than the onset times estimated based on the highest pulling rate (3 mm/min). This indicates that the faster the pulling rate the easier the generation of the shear band becomes.

The second feature seen in Fig. 1.5 is that the propagation speed of the shear band falls on a single curve as a function of the elapsed time regardless of the pulling rate. Numerical fitting indicates that this function has the form of $a/(t + b)$ where a and b are constants. It is interesting to note that the average speed of mobile dislocations τ is in the form of $\tau = L/(t + t_s)$ where L is the average distance between neighboring barrier and t_s is the interaction time of mobile dislocations with each barrier [17]. The mechanism that determines the propagation speed of shear band is not understood at this time. However, this observation strongly indicates that the shear band is formed by a physical process dependent on time and independent of pulling rate.

1.4 Conclusion

The wave dynamics of deformation was discussed using the wave characteristics observed in the differential displacement field of thin plate specimens under tensile loading. Theoretically predicted specific features in different stages of deformation have been confirmed through comparison between experimental and numerical spatiotemporal behaviors of the differential displacement field. The decrease in the propagation speed of shear band observed has been explained based on the previous interpretation that the shear band represents the solitary wave associated with the deformation dynamics in the pre-fracture stage. It has been found that the shear band propagation speed decreases as a function of the elapsed time, independent of the pulling rate of the specimen. This finding is interesting because the shear band's propagation speed itself is proportional to the pulling rate. A simple consideration indicates the possibility that the observed decrease in the shear band propagation speed is related to dynamics of mobile dislocations. Further consideration on this finding is a subject of our future study.

Acknowledgments This study was supported by the Ministry of Trade, Industry and Energy (MOTIE) and Korean Institute for Advancement of Technology (KIAT), Korea, through International Cooperative R&D program (Project No. P0006842).

References

1. Hannah, R.L., Reed, S.E. (eds.): Strain Gage Users' Handbook. Chapman and Hall, London (1992)
2. Sciammarella, C.A., Sciammarella, F.M.: Experimental Mechanics of Solids. Wiley, Hoboken (2012)
3. Sutton, M.A., Orteu, J.J., Schreier, H.W.: Image Correlation for Shape, Motion and Deformation Measurements. Springer, New York (2009)
4. Landau, L.D., Lifshitz, E.M.: Theory of Elasticity Course of Theoretical Physics, vol. 7, 3rd edn. Butterworth-Heinemann, Oxford (1986)
5. Lubliner, J.: Plasticity Theory. Courier Dover, New York (2008)
6. Irwin, G.R.: Fracture Dynamics in Fracturing of Metals. American Society for Metals, Cleveland (1948)
7. Yoshida, S.: Deformation and Fracture of Solid-State Materials – Field Theoretical Approach and Engineering Applications. Springer, New York (2015)
8. Elliott, J.P., Dawber, P.G.: Symmetry in Physics, vol. 1. Macmillan, London (1984)
9. Yoshida, S., Widiastuti, S.R., Pardede, M., Hutagalung, S., Marpaung, J.S., Muhardy, A.F., Kusnowo, A.: Direct observation of developed plastic deformation and its application to nondestructive testing. Jpn. J. Appl. Phys. **35**, L854L857 (1996)
10. Muchiar, S.Y., Muhamad, I., Widiastuti, R., Kusnowo, A.: Optical interferometric technique for deformation analysis. Opt. Exp. **2**, 516–530 (1998)
11. Yoshida, S., Siahaan, B., Pardede, M.H., Sijabat, N., Simangunsong, H., Simbolon, T., Kusnowo, A.: Observation of plastic deformation wave in a tensile-loaded aluminum-alloy. Phys. Lett. A. **251**, 54–60 (1999)
12. Nakamura, T., Sasaki, T., Yoshida, S.: Analysis of Portevin-Le Chatelier E_ect of Al-Mg alloy by electronic speckle pattern interferometry. In: Jin, H., Sciammarella, C., Yoshida, S., Lamberti, L. (eds.) Advancement of Optical Methods in Experimental Mechanics, vol. 3. pp. 109–117. Conference Proceedings of the Society for Experimental Mechanics Series. Springer, Cham (2014)
13. Sasaki, T., Nakamura, T., Yoshida, S.: Observation of grain-size E_ect in serration of aluminum alloy. In: Jin, H., Sciammarella, C., Yoshida, S., Lamberti, L. (eds.) Advancement of Optical Methods in Experimental Mechanics, vol. 3, pp. 109–115. Conference Proceedings of the Society for Experimental Mechanics Series. Springer, Cham (2015)
14. Yoshida, S., Sasaki, T.: Deformation wave theory and application to optical interferometry, Materials, in print (2020)
15. Yoshida, S.: Wave nature in deformation of solids and comprehensive description of deformation dynamics. Proc. Estonian Acad. Sci. **64**, 438–448 (2015)
16. Yoshida, S., Sasaki, T.: Field theoretical description of shear bands. In: Beese, A. M., et al. (eds.) Proceedings of SEM 2015 Annual Conference, Fracture, Fatigue, Failure and Damage Evolution 8, Chapter 18, pp. 141–149 (2016)
17. Suzuki, T., Takeuchi, S., Yoshinaga, H.: Dislocation Dynamics and Plasticity Springer Series in Material Science, 12. Springer, Berlin/Heidelberg/New York (1991)

Chapter 2
Non-contact Measurement of Strains Using Two Orthogonal Sets of Twin "Blue" Lasers

R. W. L. Fong and J. Patrick

Abstract The pressure tubes for CANDU® (CANDU® (CANada Deuterium Uranium) is a registered trademark of Atomic Energy of Canada Limited.) reactor fuel channels are made of Zr-2.5Nb alloy material. The modelling of fuel channel deformation behavior during accident scenarios, for example, a large break loss-of-coolant accident (LBLOCA), requires knowledge of the high-temperature properties of the pressure tubes. Uniaxial flat specimens are commonly used for tests to obtain the mechanical properties of the material for their response to various types of loads that simulate accident conditions. For CANDU fuel sheathing, made from Zircaloy-4 material, biaxial closed-end burst tests are usually conducted to evaluate their creep and ballooning deformation behavior at high temperatures under high heating rates. Since the Zr-alloy materials oxidize readily at elevated temperatures and their metal properties can be drastically affected, the high-temperature testing of samples from these materials should be conducted in a controlled environment, either in vacuum or surrounded by an inert gas atmosphere. By testing inside such an environment, representative mechanical properties of the metal can, therefore, be obtained with minimal effects from chemical interaction during the test.

At Canadian Nuclear Laboratories (CNL), we have developed a specially designed facility for high-temperature testing of uniaxial tensile specimens and biaxial burst specimens inside a sealed chamber. The sample is joule-heated at high heating rates with alternating current. Spot-welded thermocouple on the specimen and a PID controller are used for temperature control. Non-contact measurement of strains on the sample is made continuously, using two orthogonal sets of twin "blue" lasers. The use of blue lasers with a shorter wavelength (UV) than infrared emissions has been demonstrated to allow discernable deformations to be measured while the sample is heated to 1000 °C or higher. This paper will briefly describe the novel four-laser measurement technique with two examples to demonstrate strain measurements for the uniaxial test case and the biaxial burst test case both heated to high temperatures.

Keywords Four-laser measurement · Fuel sheathing biaxial burst · Pressure tube uniaxial test · High-temperature deformation

2.1 Introduction

The thermo-mechanical properties of zirconium-based alloy fuel sheath and pressure tube materials have to be known to properly evaluate their high-temperature response during a large break loss-of-coolant accident (LBLOCA). The mechanical testing of these materials at very high temperatures in a controlled environment using standard commercial test equipment can be restrictive and difficult.

Specialized equipment are needed to meet several test requirements. The design and operation of the test facility have to be able to test small and representative specimens over a range of high temperatures and mechanical stress conditions. An inert gas-tight chamber to enclose the specimen is required to isolate the high temperature effects on the material from the effects of chemical interactions (e.g., heavy oxidation or hydriding). The specimen temperature and high heating rates and stress loading are controlled to simulate accident conditions. Strain measurements at high temperatures are required with online acquisition capability without affecting the mechanical behavior of the specimen during the test.

Burst behavior of fuel sheath under various accident scenarios has been extensively studied. Under some accident conditions, the fuel sheath deforms as a result of internal pressure and high temperature excursion that can lead to ballooning

R. W. L. Fong (✉) · J. Patrick
Canadian Nuclear Laboratories, Chalk River, ON, Canada
e-mail: randy.fong@cnl.ca; james.patrick@cnl.ca

M.-T. Lin et al. (eds.), *Advancement of Optical Methods & Digital Image Correlation in Experimental Mechanics*, Conference Proceedings of the Society for Experimental Mechanics Series, https://doi.org/10.1007/978-3-030-59773-3_2

(extensive hoop strain) before the tube bursts. If the tube is highly stressed axially, in a certain biaxial stress state, contractile deformation (inverse of ballooning) can result as the tube elongates in length and breaks apart, axially. A strong anisotropy in creep behavior is evident in zirconium-based alloy materials that is highly correlated to the preferred orientation of α-phase grains in the microstructure. The deformation behavior (e.g., creep rates) is also affected by oxidation or hydriding. For pressure tube materials made from Zr-2.5Nb alloy (a dual-phase material), the high-temperature deformation behavior is consistent with that of fuel sheath single-phase materials made from dilute zirconium alloys (e.g., Zircaloy-2, Zircaloy-4, or Zr-1Nb).

At Canadian Nuclear Laboratories (CNL), we have developed a specially designed facility for thermo-mechanical testing of uniaxial tensile and fuel cladding burst specimen for test inside a gas-tight environmental chamber. The facility is equipped with a uniaxial tensile/compressive loading system, a pressurizing gas delivery system, and an AC power supply for resistive heating of the specimen. In particular, this test facility is specially equipped with a four-laser scanning system for non-contact strain measurement. Techniques such as using video extensometer [1], mechanical LVDT extensometer [2, 3], optical telescope [4], photographs [5], high-powered digital cameras [6, 7], bonded strain gauges [8, 9], and laser extensometer [10–14] have been used, but all these methods can only measure a one-directional (azimuthal) diameter of the tube which may not be its maximum diameter. Therefore, we have developed a 4-laser measurement technique that allows for online scanning of the fuel sheath specimen during the test. This technique provides a means to determine the maximum deformation rate of change in the specimen. In biaxial burst testing of fuel sheath, the changes in maximum creep rate during creep and ballooning usually corresponds to the tube maximum bulge (diameter) location, where metal instability would first occur. As such, representative time events of the deformation data can now be extracted from the online scanning measurement. The representative data of the tube's deformation behavior can be used for modelling or testing existing deformation equations.

This paper provides a brief description of design features of CNL's biaxial burst test facility developed for thermo-mechanical testing of fuel sheath and uniaxial tensile specimens. Two examples of non-contact strain measurements made on these two types of specimens are presented.

2.2 Experimental Design

2.2.1 Design of CNL's Online Biaxial Burst Test Facility

Figure 2.1 shows a schematic diagram of the design of CNL's biaxial burst test facility. A specially designed gas-tight chamber is mounted on a commercial tensile testing machine, which provides a controlled uniaxial tensile or compressive axial stress loading on a test specimen.

Although not illustrated in the figure, the chamber has three mutually accessible compartments (top, middle, and bottom boxes), where the top and bottom boxes housed the feedthroughs for electrical buss bars for connections of an AC power supply to the specimen. The sample is joule-heated at high heating rates with alternating current. A spot-welded thermocouple on the specimen and a PID controller are used for temperature control.

The bottom box is equipped with access ports for gas evacuation and for purging (back filling) argon and a port for the pressuring gas inlet line connection to the specimen for burst testing. The gas lines are electrically isolated using ceramic spacers. A rupture disk is equipped in the bottom box for rapid exhaustion of excess pressurized gas from the chamber into an expansion tank immediately when a tube sample bursts.

The top box is equipped with a flexible gas-tight (bellows) connection to the middle box. It has an access port for the pressurizing gas outlet line connected to the top end of the burst specimen that is used for argon purging before test and for bleeding the pressurizing gas for control of internal pressure in the burst specimen during the test. A pressure transducer installed at the outlet line is used to measure the pressure inside the fuel sheath burst sample. The top of this box is connected to an additional compartment equipped with feedthroughs for thermocouple wire sensors, pressure transducer, piping lines for cooling-jackets, and a flexible gas-tight (bellows) connected to the pull bar of the tensile testing machine.

The middle box surrounding the test specimen has eight-sided walls, which is equipped with two demountable covers secured on opposite sides of the chamber. The two covers both have two gas-tight quartz windows that provide four cardinal directions for viewing with the four lasers used to scan the sample. One of the demountable covers is used to allow access for changing the test specimen on the tensile machine, installing spot-welded thermocouples on the specimen, connection of the pressuring gas inlet and outlet lines (for bursting the sample), and sample rotation alignment to the laser beam (mainly for a flat specimen) before closing the cover.

Four-laser displacement sensors (KEYENCE model LJ-V7200, "blue" line laser), mounted on two mechanical slides, are positioned outside of the octagonal chamber. The slides are programmed to move the lasers up and down to allow axial

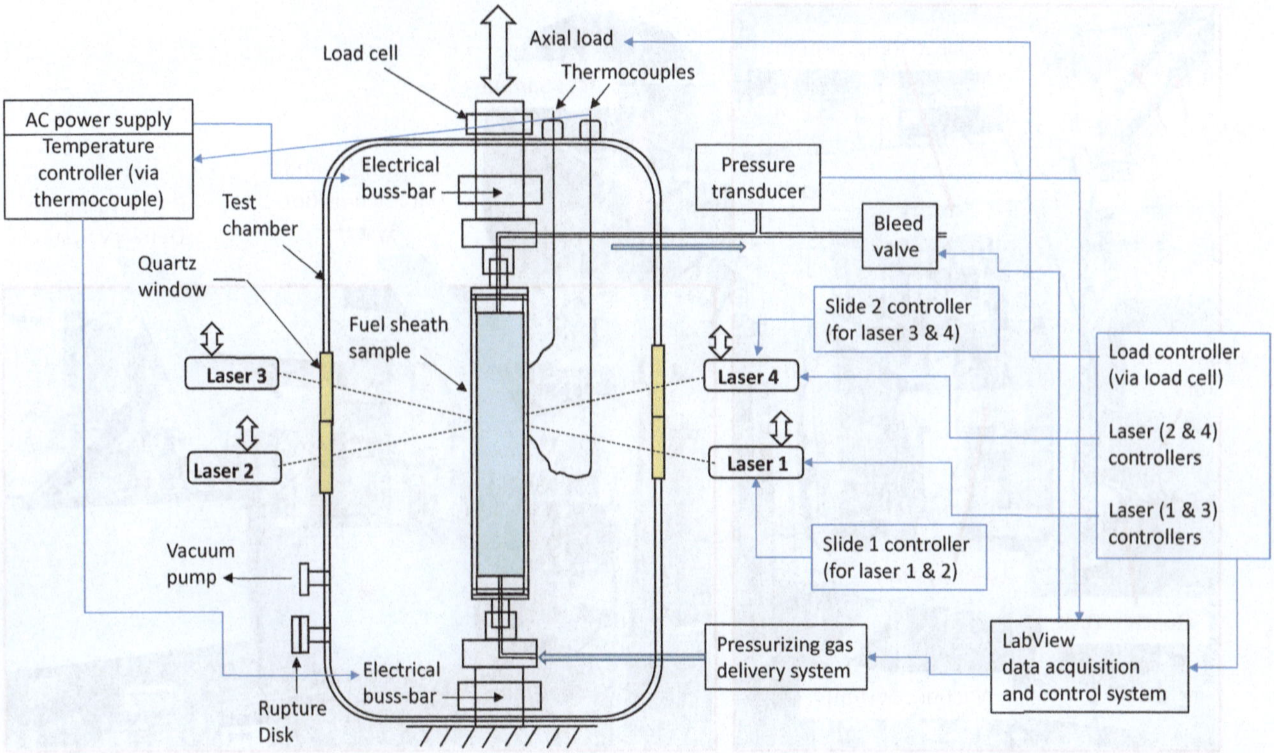

Fig. 2.1 Schematic diagram of CNL's online biaxial burst test facility using four lasers for non-contact strain measurement

scanning of the specimen. Both slides are equipped with LVDTs to obtain the axial (vertical) positions of the lasers. The use of "blue" lasers (with a UV wavelength outside of the infrared regime) allows displacement measurements to be made on a sample heated to high temperatures up to 1000 °C. Each laser is aligned using a U-joint and X–Y rotation stage to be set along a targeted cardinal direction. All four lasers are used to scan the specimen for strain measurements, each laser measuring more than one quadrant of the specimen. With the use of two orthogonal sets of twin lasers, the total circumference of the fuel sheath can be measured during creep and ballooning continuously without contacting the specimen. Similarly, the width and thickness dimensions of a flat specimen are measured without contact. Two video cameras, positioned at the front and back quartz windows, are used to record the sample ballooning and burst during the test.

A LABVIEW data acquisition and control system is used to regulate the alternating trigger duty cycle of the two sets of twin lasers and collect all the laser displacement data. Also simultaneously, data are collected from the pressure transducers (for internal pressure in the specimen and test chamber), thermocouples (for sample temperatures), tensile machine (for axial load and displacement on the specimen), and the two slides (LVDTs positions). Figure 2.2 shows photographs of the complete online biaxial burst test facility for thermo-mechanical testing of fuel sheath and flat tensile samples.

2.2.2 *Test Specimens*

Figure 2.3 shows two types of test specimens: (a) burst specimen and (b) flat tensile specimen. The burst specimen (Fig. 2.3a) is a tube, with the typical application being a sample of fuel sheath, which is sealed with two laser-welded end plugs that have a hollow spigot at the ends equipped with Swagelok fittings for connection to the pressurizing gas inlet and outlet lines. Specimens 130 mm in length and 9–15 mm outside diameter, typical of fuel sheath samples, can be investigated. Changes in the average tube circumference displacement as small as 0.015 mm can be resolved with this method.

Flat (dog-bone) tensile specimen (Fig. 2.3b) can be tested. One typical application is a transverse specimen, cut from the transverse direction of a pressure tube material. The specimen has a gauge length of 10 mm, 6 mm in width, and 3 mm in wall thickness. Changes in the average displacements of the width and thickness dimensions as small as 0.005 mm can be resolved.

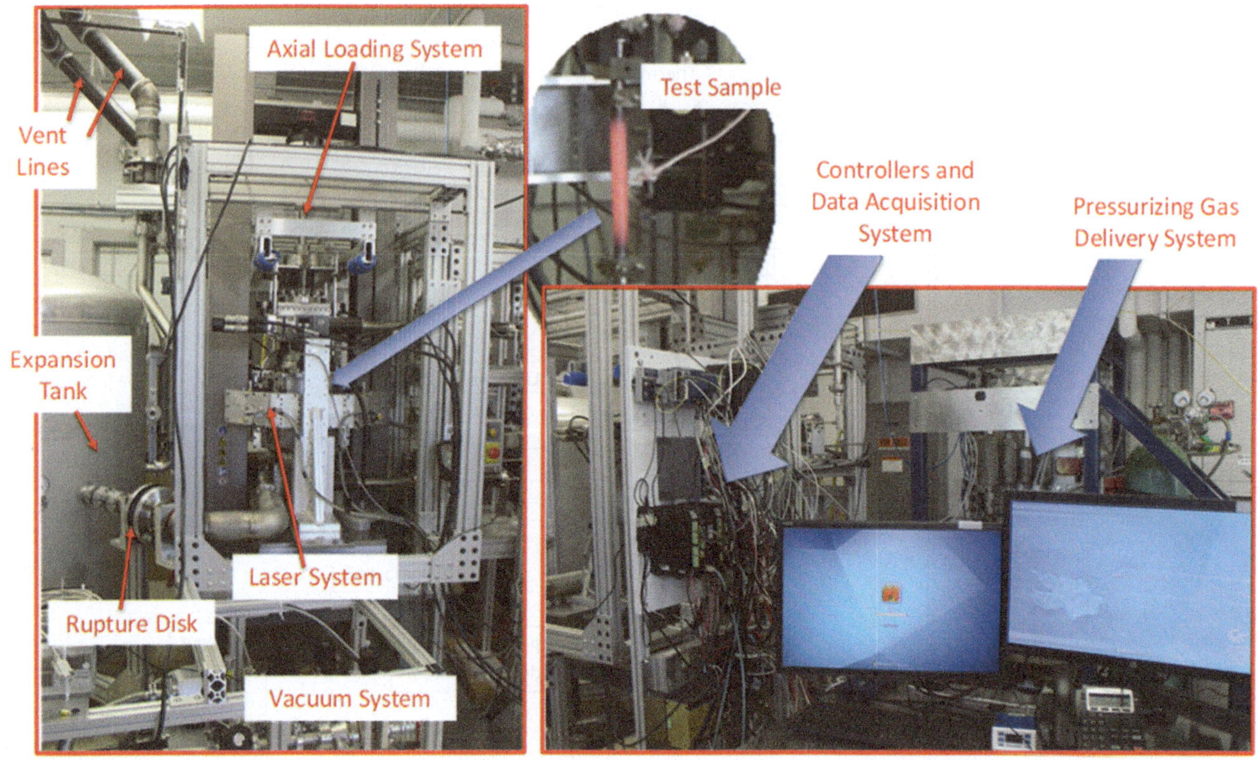

Fig. 2.2 Photographs showing the online biaxial burst test facility at CNL

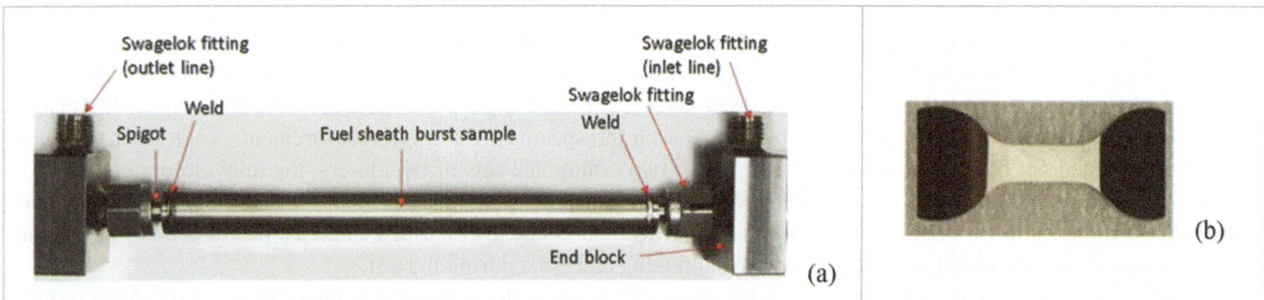

Fig. 2.3 (**a**) Fuel sheath burst specimen, (**b**) flat (dog-bone) uniaxial tensile specimen

2.2.3 Test Procedure

For biaxial burst test of a fuel sheath sample, a controlled axial loading on the tube allows different magnitudes of axial to hoop stressing ratios to be applied to deform the tube during pressurization. The different stress states (or end restraints) are useful to study the effects on ballooning deformation. For testing a flat tensile specimen, the axial loading applied will simply generate a uniaxial stress to elongate the specimen along the loading direction.

The following preparation is typically used to setup a fuel sheath sample ready for biaxial burst test.

1. Open the cover of test chamber and install a test sample on the tensile machine.
2. Install the pressuring gas lines to the sample inlet and outlet fittings.
3. Spot weld thermocouples and a Zr tab on the sample.
4. Check for electrical continuity on the sample and electrical isolation from the remainder of the test apparatus.
5. Close the cover, and install a rupture disk on the bottom box.
6. Evacuate the test chamber and purge (backfill) with argon, and repeat this step several times.
7. Purge the inside of the sample with argon and close the bleed valve.

Once the sample is readied, a burst test comprises the following steps.

1. Apply a prescribed controlled constant axial load on the sample using the tensile machine. This allows a prescribed biaxial stress state to be imposed on the tube sample (e.g., to simulate a "closed-end" type of burst test where the axial stress is half the hoop stress – i.e., axial-to-hoop stress ratio of 0.5).
2. Enable all the controllers, vis-à-vis, (a) four lasers, (b) two slides, (c) PID temperature feedback for AC power supply, (d) video cameras, and (e) LABVIEW data acquisition and control system (programmed to control alternating trigger cycling duty of the two orthogonal set of twin lasers, and controlling motorized needle valves for the sample pressuring gas feed and bleed system).
3. Start the slides which holds the four lasers to move repeatedly up and down to scan the sample to a set distance above and below the spot-welded Zr tab. If desired, the slides can be set at a fixed (stationary) axial location to scan the sample.
4. Start the LABVIEW program.
5. Heat the sample using PID controls to prescribed test conditions. The prescribed test conditions can be a simple single creep test (at a desired constant temperature) or staged creep test (multiple steps of different constant temperatures), a ramped pressure, or a ramped temperature test condition.
6. Once the test is completed, allow the sample to cool to room temperature then obtain laser scans on the burst sample at various axial locations and at the maximum ruptured diameter location, before removing the sample for ex situ post-test examination.
7. Disable all controllers, and close out the LABVIEW program.

2.3 Analysis

The analysis of displacement data collected by the four lasers at a given time is processed using a MATLAB routine which we have developed to determine the full circumference of the tube that is used to calculate the hoop strain. The hoop strain is calculated using the following relation: hoop strain (%) = $(c_i - c_o) \times 100/c_o$, where c_i = instantaneous tube circumference and c_o = original (initial) tube circumference. The value for total circumferential elongation (TCE) is obtained using the relation: TCE (%) = $(c_f - c_o) \times 100/c_o$, where c_f = final tube circumference and c_o = original (initial) tube circumference.

Figure 2.4 illustrates an example of an alternating duty cycle of the two sets of twin lasers (laser 1 and 3 and laser 2 and 4), the individual displacement profiles measured, and the analyzed tube circumference determined using the MATLAB routine. Using an alternating duty cycle as shown in Fig. 2.4a, b, we have found that this measurement technique clearly eliminates crossover speckle interference originating from immediate adjacent (orthogonal) lasers. With the twin lasers 1 and 3 (diametrically opposite each other) operated from the same Keyence controller, the displacement profile of laser 1 is flipped about its y-axis and has positive values for y-displacement, and for laser 3, the profile is not flipped about its y-axis but has negative values for the y-displacement (Fig. 2.4c). The characteristics of the y-displacement profiles and orientations produced by the lasers as configured in the Keyence control operation have been verified with scans made on a four-sided calibration block with distinctly different profiles machined on each of the four faces. Similarly, for the twin lasers 2 and 4 which operated using a different Keyence controller, the y-displacement profiles and orientations produced are of the same characteristics as those generated by the twin lasers 1 and 3 (Fig. 2.4d).

To determine the tube full circumference at a given time (Fig. 2.4e), the MATLAB routine follows the steps below to process the displacement profiles generated by the four lasers. Figure 2.5 illustrates the process step (1–8) listed below.

1. Flip the scan line (raw x–y data) about the y-axis for laser 1 and laser 2 (Step 1).
2. Trim "out-of-range" (y-displacement) data points, and remove excess data points at both tails if the length of the line is too long (Step 2).
3. Level the trimmed line to be parallel to the x-axis (Step 3).
4. Translate the levelled line with its mid-point centered on the centerline axis and its y-value corrected by adjustment to the untested tube radius (Step 4).
5. Rotate the line to the respective quadrant (e.g., laser 1 line to quadrant 1, and so on) (Step 5).
6. Fit a circle to the line to find the center x–y coordinates and the radius of the fitted circle. Using the known diameter of the tube before test, the levelled line is adjusted by calibrating to the known radius of the untested tube (Step 6).
7. Using the fitted center coordinates, the line is translated to the origin (0, 0) to be coincident with the axis of the tube, to be in the sample coordinates system (Step 7).
8. Trim each line along the x and y axes so that the line is solely contained within its respective quadrant (Step 8).

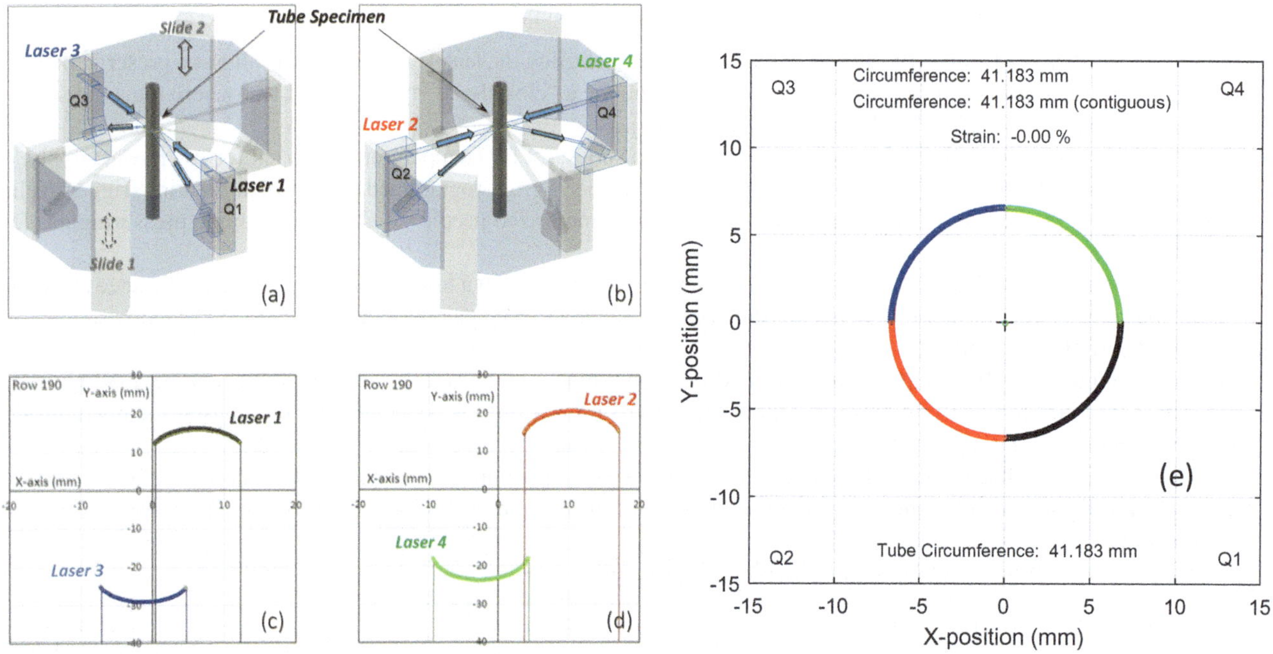

Fig. 2.4 Two sets of twin lasers measurement system: (**a**) twin lasers 1 and 3 at Q1 and Q3: on duty (while orthogonal twin lasers 2 and 4 at Q2 and Q4: off duty; (**b**) twin lasers 2 and 4 at Q2 and Q4: on duty (while orthogonal twin lasers 1 & 3 at Q1 and Q3: off duty; (**c**) profiles measured by lasers 1 & 3; (**d**) profiles measured by lasers 2 & 4; and (**e**) analyzed complete profile of tube circumference obtained from the 4 orthogonal lasers

9. All four lines are organized into a contiguous string of (x–y) data points to calculate the tube circumference at that given time. This instantaneous tube circumference (c_i) is obtained by summing all the lengths between all two adjacent data points in the entire contiguous string of data points (Fig. 2.4e).
10. The instantaneous tube circumference (c_i) is used to calculate the tube hoop strain using the relation given earlier (Fig. 2.4e).

2.3.1 Example of Analysis of a Fuel Sheath Biaxial Burst Specimen

As mentioned earlier, the ability to scan the sample at different axial locations during the burst test provide a useful means to be now able to extract the maximum strain rate corresponding to where the maximum bulge (ballooning) strain is occurring on the tube as it progressively deforms to rupture. This measurement technique generates precise information on the onset event of metal instability and deformation behavior. In addition, since this four-laser measurement technique can capture the entire profile of the tube circumference, there is no ambiguity to discern whether or not the maximum diametral strain would have been captured.

Figure 2.6 presents an example of analysis using the four-laser measurement technique where the tube is scanned all around the tube circumference at each different axial location on the tube by sliding all four lasers a short distance up and down (above and below) the Zr-tab which is used as reference position marker. From experience, the spot-welded Zr-tab is placed on the tube at a location fairly near where the maximum (bulge) ballooning is be expected to occur. The photograph shown on the left hand side of Fig. 2.6 shows the Zr-tab on the fuel sheath burst sample. On the right-hand side of the same figure are the results of analysis of the tube circumference (with no distorted circumferential profile) for the locations just above and below the Zr-tab for two cases. In Case A, the lasers moved upward, and Case B when the lasers returned and moved downward. At the Zr-tab locations, the laser scan results show the tube circumferential profile to be distorted, thus providing an indication of the reference marker location on the tube.

Figure 2.7 shows some results of the four-laser measurement of the full tube circumference of a fuel sheath sample during a ballooning test to burst. The snapshot photographs taken from the video camera show the axial location of the laser scan on the tube relative to the location of the Zr-tab reference marker.

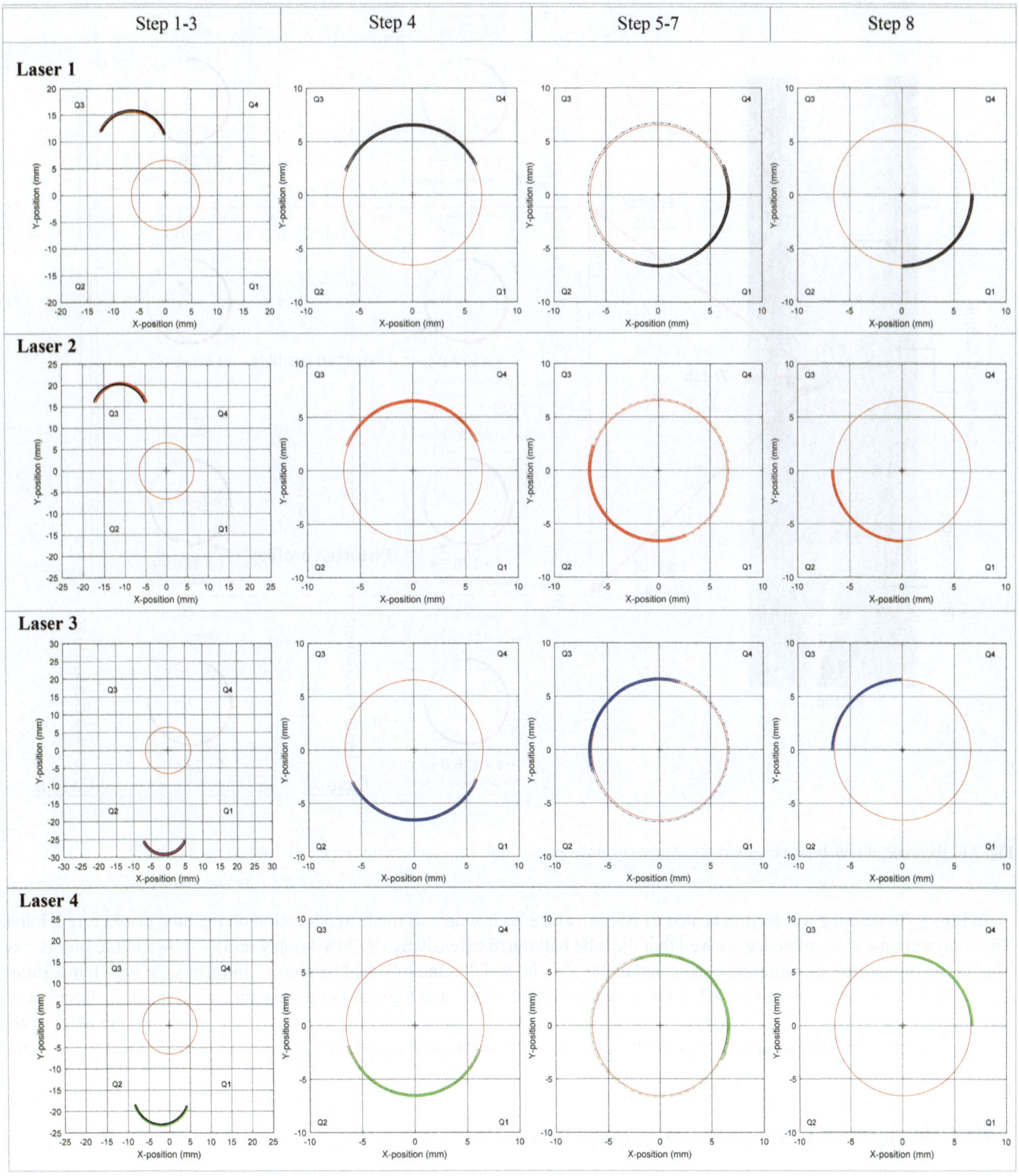

Fig. 2.5 Processing steps in MATLAB routine used to determine the tube full circumference from four-laser measurement

In Fig. 2.7a, when the tube was heated at 600 °C, the results of analysis of the tube circumference as shown in the plot below showed that some creep strain (~4% of hoop strain) was detected and the profile of the circumference remained circular. A circular profile suggests that the creep deformation occurred uniformly all around the circumferential location of the tube.

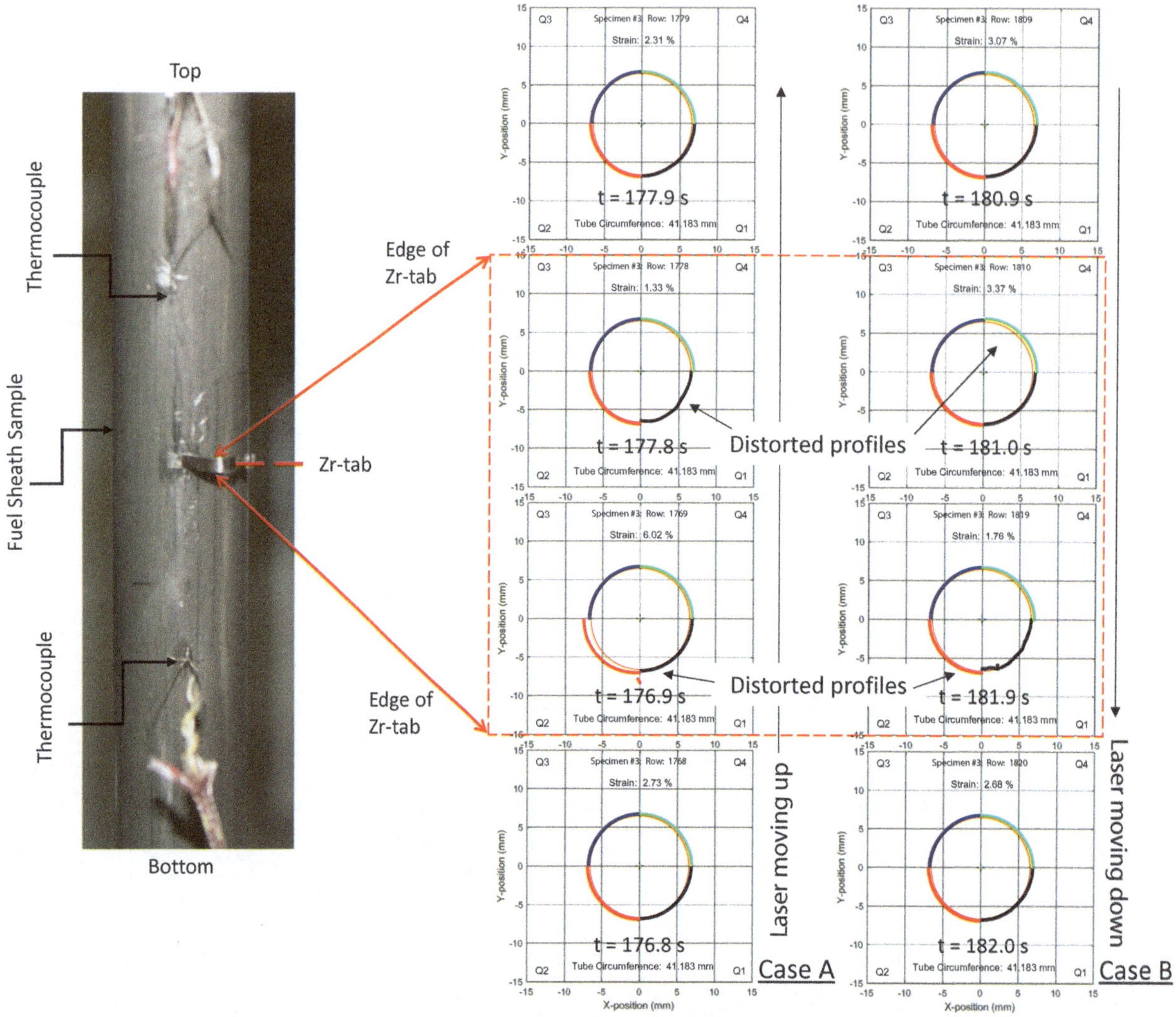

Fig. 2.6 Example of four-laser measurements of tube circumference at different axial location on the fuel sheath burst sample

In Fig. 2.7b, the tube was further heated to 700 °C, and a laser scan was made at a location on the tube just before it burst. The analyzed tube circumference showed that the tube ballooned extensively (>120% hoop strain) – shown in the plot below. The profile of the tube circumference indicated in the plot showed that the tube had ballooned uniformly. At high temperature uniform ballooning is usually expected, whereas at low temperatures or at high pressures, asymmetric (lopsided) deformation and burst can be expected. This translates to increasing the difference between the maximum and minimum diameter of the tube. As such, the measurement results could be affected or biased depending where the diameter measurement was taken.

2.3.2 Example of Analysis of a Dog-Bone Specimen

A slightly different MATLAB routine is used for analysis of the laser data for a flat sample as it has a four-sided cross-section geometry. The displacement data collected by the four lasers in the width and thickness directions are used to compute the tensile axial strain in the specimen (i.e., along the loading direction). This MATLAB routine does not require the above Steps 6 and 7 (i.e., fitting a circle to the line profile and using the fitted center to translate the profile to the origin of the sample coordinates). These two processing steps are replaced by simply pre-determining the standoff distance of each laser sensor and calibrating the standoff value to translate the line profile that fits the sample dimension in sample coordinates. The following relation is used to calculate the true strain in each principal direction: $\varepsilon_{\text{true}} = \ln(l/l_o)$, where l and l_o are the instantaneous dimension and the original dimension in that direction, respectively. Assuming constant volume, the strain in

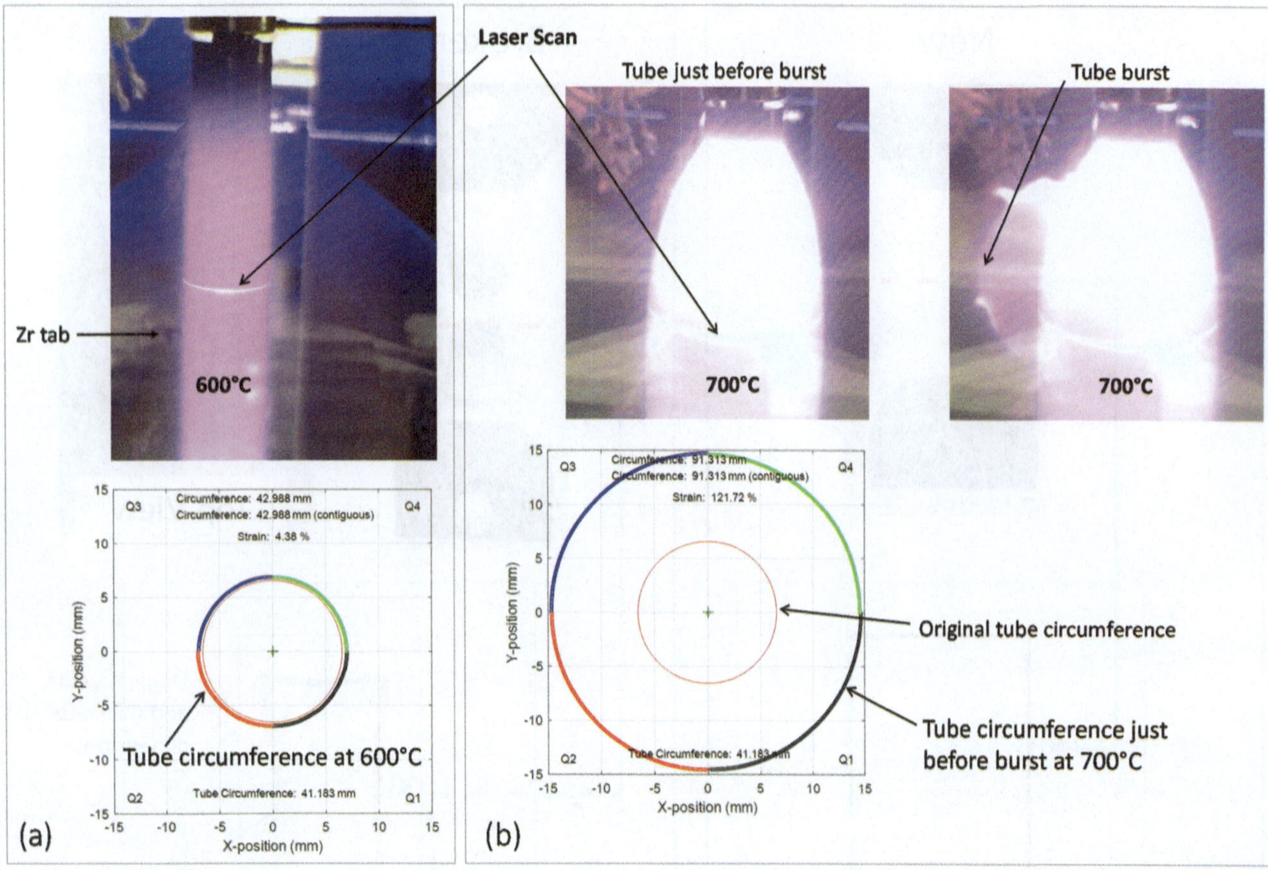

Fig. 2.7 Example of four-laser measurements of tube circumference of a fuel sheath sample tested at 600 °C and 700 °C (just before tube burst)

the tensile direction is then: $\varepsilon = -(\varepsilon_W + \varepsilon_t)$, where ε_W and ε_t are strains in the width and thickness direction of the sample, respectively.

Figure 2.8 shows an example of results of analysis of displacement data measured by four lasers on a flat tensile specimen. In Fig. 2.8a, the four-laser line profiles which have been calibrated using known dimensions (width and thickness) of the new specimen is given in the plot below. Using the calibrated parameters set in the MATLAB routine, the necked dimensions and profile at a particular location on the broken sample can then be precisely analyzed as shown in the plot below in Fig. 2.8b. The laser measured profile is shown to closely match the boundary profile of the fractured surface of the broken samples (top view).

2.4 Conclusion

The CNL's biaxial burst test facility for thermo-mechanical testing at high temperatures has been briefly described. The four-laser scanning system equipped in this facility has been developed and used successfully for online non-contact measurement of hoop strains in a fuel sheath specimen during creep and ballooning deformation during biaxial burst testing. The four-laser scanning system has also been successfully developed to gather multiple strain measurements at different axial locations of tube that provides an "in situ" means to capture the onset and maximum creep rate during creep and ballooning up until burst at high temperatures. The four-laser scanning technique has also been used successfully for strain measurement on uniaxial flat tensile specimens.

Acknowledgments This development and experimental work was funded by Atomic Energy of Canada Limited, under the auspices of the Federal Nuclear Science and Technology Program.

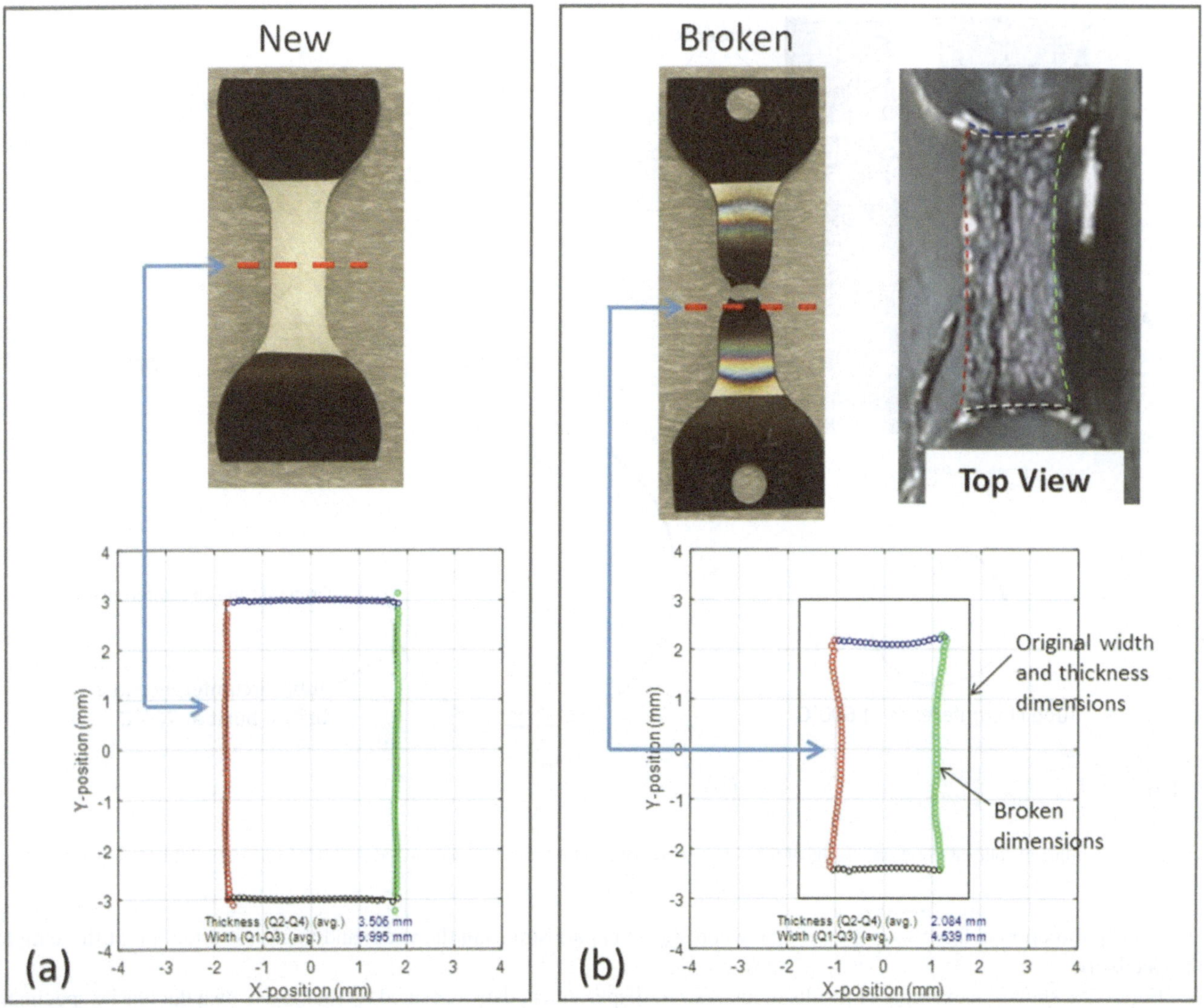

Fig. 2.8 Example of four-laser measurements of new and broken tensile specimen

References

1. Völkl, R., Fischer, B.: Mechanical testing of ultra-high temperature alloys. Exp. Mech. **44**(2), 121–127 (2004)
2. Yadav, A.K., Shin, C.H., Lee, S.U., Kim, H.C.: Experimental and numerical investigation on thermo-mechanical behavior of fuel rod under simulated LOCA conditions. Nucl. Eng. Design. **337**, 51–65 (2018)
3. Yadav, A.K., Shin, C.H., Lee, C., Lee, S.U., Kim, H.C.: Experimental investigations on out-of-pile single rod test using fuel simulator and assessment of FRAPTRAN 2.0 ballooning model. Ann. Nucl. Energy. **124**, 234–244 (2019)
4. Nagy, R., Király, M., Szepesi, T., Nagy, A., Almási, A.: Optical measurement of the high temperature ballooning of nuclear fuel claddings. R. of Sci. Instr. **89**, 125114, 1–7 (2018)
5. Suzuki, M.: High temperature deformation behavior of gradually pressurized zircaloy-4 tubes. J. Nucl. Sci Tech. **18**(8), 617–628 (1981)
6. Gussev, M.N., Byun, T.S., Yamamoto, Y., Maloy, S.A., Terrani, K.A.: In-situ tube burst testing and high-temperature deformation behavior of candidate materials for accident tolerant fuel cladding. J. Nucl. Mater. **466**, 417–425 (2015)
7. Cinbiz, M.N., Gussev, M., Linton, K., Terrani, K.A.: An advanced experimental design for modified burst testing of nuclear fuel cladding materials during transient loading. Ann. Nucl. Energy. **127**, 30–38 (2019)
8. Cinbiz, M.N., Brown, N.R., Terrani, K.A., Lowden, R.R., Erdman III, D.: A pulse-controlled modified-burst test instrument for accident-tolerant fuel cladding. Ann. Nucl. Energy. **109**, 396–404 (2017)
9. Li, F., Mihara, T., Udagawa, Y., Amaya, M.: Biaxial-EDC test attempts with pre-cracked zircaloy-4 cladding tubes. Proc. ICONE25 (2017)
10. Grosjean, C., Poquillon, D., Salabura, J.-C., Cloué, J.-M.: Cladding tube testing in creep conditions under multiaxial loadings: a new device and some experimental results. Trans., SMiRT 19, Paper # C01/2, Toronto (2007)
11. Inoue, T., Ogawa, R., Akasaka, N., Nishinoiri, K.: Transient burst techniques and results of the examination for irradiated PNC316 steel,, JAEA-Conference 2008–010, KAERI/GP-279/2008 (2008)

12. Grosjean, C., Poquillon, D., Salabura, J.-C., Cloué, J.-M.: Experimental creep behavior determination of cladding tube materials under multi-axial loadings. Mater. Sci. Engr. A. **510–511**, 332–336 (2009)
13. Priser, M., Rautenberg, M., Cloué, J.-M., Pilvin, P., Feaugas, X., Poquillon, D.: Multiscale analysis of viscoplastic behavior of recrystallized zircaloy-4 at 400 °C. J. ASTM Inter., Paper ID JAI103015. **8**(1), 1–19 (2011)
14. Yueh, K., Karlsson, J., Stjärnsäter, J., Schrire, D., Ledergerber, G., Munoz-Reja, C., Hallstadius, L.: Fuel cladding behavior under rapid loading conditions. J. Nucl. Mater. **469**, 177–186 (2016)

Chapter 3
Experimental Observations on the Fracture of Metals

C. A. Sciammarella, L. Lamberti, and F. M. Sciammarella

Abstract This paper deals with the onset of plasticity and the transition to fracture in metallic rectangular tensile specimens. The plastic instability manifests itself by the propagation of wave fronts that sweep the specimen and finally localize at a given site during the fracture process. The plastic instability is associated with changes in specimen geometry. In optical recordings, propagation and localization of structural instability manifest in the form of wide bands. The configuration of these bands is influenced by degradation of properties of a metal under analysis. Understanding the meaning of band configurations is vital for evaluating mechanical properties of a metal and its effective use. In this paper, a family of fringes called iso-derivatives is utilized to analyze the configuration and properties of the wide bands. The retrieval of the information contained in the wide bands depends on the spatial and temporal resolution of the recordings.

Keywords Plasticity · Fracture mechanics · Metals · Optical methods

3.1 Introduction

The process of failure in materials is a complex problem that present challenges from a fundamental point of view, full understanding of diverse mechanisms leading to the instability of dynamic equilibrium in loaded solids. In spite of theoretical and experimental developments, actual dynamics of a fracturing solid, connection between motion equations with molecular structure of the solid, presents many unknown aspects. Basic questions that theoretical efforts of atomistic based models, as well as the Continuum approach try to answer, have deep practical consequences for scientists and engineers. Which are the mechanisms of atomic configurations or of the parameters of the Continuum approach that make it possible to predict the onset of plasticity in a given material under a variety of externally applied loadings? This question is followed by another question: transition of the onset of plasticity to failure and fracture. Currently, computer science, computing hardware, and numerical methods for solving partial differential equations provide the means to solve complex theoretical models of molecular dynamics. It is possible to summarize the current state-of-the-art of the partial obtained answers. The problem of stability of crystalline arrays is formulated in terms of the stability of internal energy W expressed as a function of suitable tensorial forms of displacement gradient field **F** and a corresponding form of a stress tensor **P**. In the case of metals, onset of plasticity implies the motion of dislocations through crystalline array. This is the approach of dynamic time-dependent plasticity evolved in Materials Science. There is another important event, the transition from plasticity to fracture. From the engineering point of view, this transition is handled by Fracture Mechanics. The near field at the tip of a crack is modeled by an approximate solution of the theory of elasticity that assumes the existence of a singularity where stresses become infinite. The extensions of this approach including the presence of a plastic zone at the crack tip also assume the existence of a singularity at crack tip. This approach leads to the formulation of transition from plasticity to fracture by introducing

C. A. Sciammarella (✉)
Department of Mechanical, Materials and Aerospace Engineering, Illinois Institute of Technology, Chicago, IL, USA
e-mail: sciammarella@iit.edu

L. Lamberti
Dipartimento Meccanica, Matematica e Management, Politecnico di Bari, Bari, Italy
e-mail: luciano.lamberti@poliba.it

F. M. Sciammarella
MXD Corporation, Chicago, IL, USA
e-mail: federico.sciammarella@mxdusa.org

M.-T. Lin et al. (eds.), *Advancement of Optical Methods & Digital Image Correlation in Experimental Mechanics*,
Conference Proceedings of the Society for Experimental Mechanics Series, https://doi.org/10.1007/978-3-030-59773-3_3

the J-integral formulation. In both elastic and elastoplastic approaches, instability of the crack propagation is based on the balance of the elastic energy stored in the material and the energy required to form new crack surfaces. In the approach of the atomistic-quasi-continuum mechanics, transitions from plasticity to fracture are also formulated on the basis of the energy analysis.

There is an alternative approach to the formulation of the onset of plasticity and of the transition from plasticity to fracture based on the gauge theoretical formalism of physical mesomechanics. Based on an initial work of V.E. Panin [1], it has been further developed and experimentally supported by the work of S. Yoshida and his collaborators. In this approach, the relationship between materials science and continuum kinematics is based on the utilization of Maxwell equations to formulate the relationship of the microstructure with the kinematics of the continuum. A very interesting yield of this method is the experimental tool utilized to connect theoretical predictions with experimental observations [2–6]. This tool is termed as cine-speckle interferometry and is an independent discovery of a family of fringes initially introduced by A.J. Durelli and V.J. Parks [7]. These fringes are space derivatives of the isothetic lines (moiré fringes). Additional applications of this family of fringes can be found in [8–11].

The present study attempts to reformulate this approach fostering a closer connection between materials sciences and continuum kinematics. The final goal is to develop a practical tool to experimentally determine the transitions to plasticity and from plasticity to fracture in metals. These transitions are dynamic events that can be observed via optical methods of Experimental Mechanics. This approach opens the possibility of establishing the connection of microstructure and continuum mechanics predictions on quantitative basis. The developments presented in this paper are restricted to metals and concentrate on tensile tests results.

3.2 Analysis of the Displacement Field of a Tensile Specimen

Let us review some fundamental aspects of the kinematics of tensile specimens. One of the manifestations of the onset of plasticity in metals from the Experimental Mechanics point of view is the change in shape of projected displacements fringe patterns, isothetic lines, or moiré fringes. In the case of materials like crystalline metals, the linear relationship between applied loads and fringe patterns implies that the geometry of the isothetic lines is preserved in the following sense. The distribution of displacements and associated loads remains proportional, or also one can say displacements and loads are linearly correlated. Increasing the load, the displacements increase in the same proportion. As a consequence of this proportionality, the ratios of displacements for successive loadings at a given point remain constant throughout the entire body, and this ratio will be in the same proportion of the ratio of the applied loadings but different for different points. Since isothetic lines represent loci of equal projected displacements and the gradients reflect changes in these projected displacements, the resulting configuration of fringes by increasing the load will reflect this constant ratio for the entire region under observation. A simple example, a tensile specimen of constant section subjected to axial load in the elastic range, will be analyzed in what follows (Fig. 3.1).

According to the basic law of isothetic lines, the displacement vectors **u** and **v** have moduli,

$$|\mathrm{u}| = p \tag{3.1}$$

$$|\mathrm{v}| = p \tag{3.2}$$

In Eqs. (3.1) and (3.2), p is the pitch of the printed cross-grating on the surface of a tensile specimen.

The modulus of the gradient in the x-direction can be computed by the approximate expression,

$$\varepsilon_x = \frac{\partial u}{\partial x} \approx \frac{p}{\delta_u} \tag{3.3}$$

And in the y-direction,

$$\varepsilon_y = \frac{\partial \mathrm{v}}{\partial y} \approx \frac{p}{\delta_\mathrm{v}} \tag{3.4}$$

The displacement vector is

$$\mathbf{d} = \mathrm{u}\,\mathbf{i} + \mathrm{v}\mathbf{j} \tag{3.5}$$

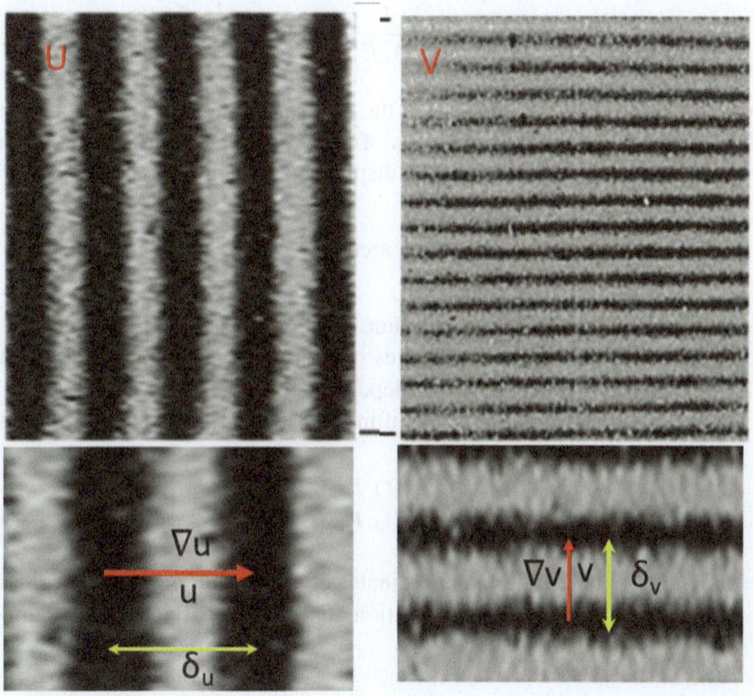

Fig. 3.1 U and V patterns of a tensile specimen with vertical and horizontal gratings of pitch p

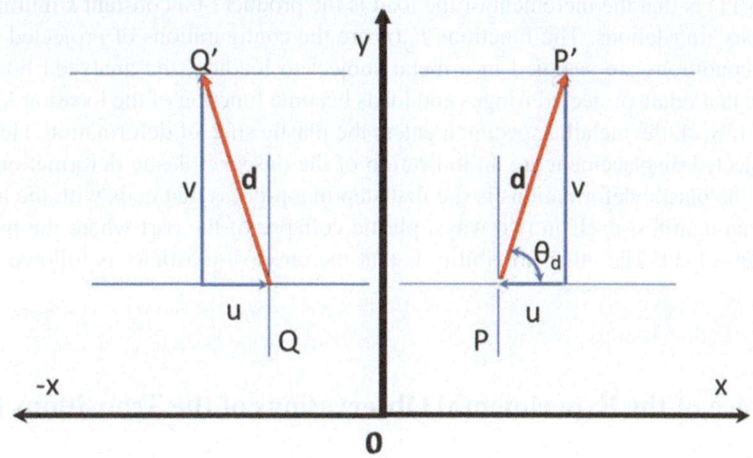

Fig. 3.2 Vector displacements in a tensile specimen of constant cross section

Figure 3.2 represents the vectors of relative displacements if the reference axis is taken at the center of the specimen. The specimen elongates in the y-direction and contracts in the x-direction due to the Poisson's effect.

In the elastic range, the corresponding values of the displacement components are

$$\mathrm{v} \approx \frac{\sigma_y}{E}\delta_{\mathrm{v}} \tag{3.6}$$

$$\mathrm{u} \approx -\nu\frac{\sigma_y}{E}\delta_u \tag{3.7}$$

In Eq. (3.7), ν is the Poisson's ratio.

The modulus of the displacement vector is

$$|\mathbf{d}| = \sqrt{u^2 + v^2} = \frac{P}{A_c E}\left(\delta_v^2 + \nu^2 \delta_u^2\right)^{\frac{1}{2}} = KP \tag{3.8}$$

In Eq. (3.8), P is the load applied to the specimen, A_c is the cross section of the specimen, E is the modulus of elasticity of the specimen, and ν is the Poisson's ratio. The modulus of the vector is proportional to the applied load, where K is a constant for the whole field. The angle of inclination of the displacement vector is

$$\theta_d = \text{arctg}\frac{\delta_v}{\nu \delta_u} \tag{3.9}$$

The observed patterns are affine transformations of the same configuration and will remain this way until the metal yields. Then, yielding is characterized by the configuration changes of moiré fringes of projected displacement. This conclusion is true for any type of fringe patterns that are linearly dependent on the applied loads. The above conclusion can be mathematically expressed by the following conditions. An affine transformation is defined in 2D as

$$\begin{bmatrix} x_1 (X_1, X_2, t) \\ x_2 (X_1, X_2, t) \end{bmatrix} = \begin{bmatrix} F_{11}(t) \; F_{12}(t) \\ F_{21}(t) \; F_{22}(t) \end{bmatrix} \begin{bmatrix} X_1 \\ X_2 \end{bmatrix} + \begin{bmatrix} c_{1T}(t) \\ c_{2T}(t) \end{bmatrix} \tag{3.10}$$

In this equation, x_i ($i = 1,2$) represent the Eulerian coordinates of a 2D medium in the deformed position, X_i represent the Lagrangian coordinates of the point in the undeformed position, and "t" is a parameter that can be the time but is related to the load, by

$$\Delta P = k \times t \tag{3.11}$$

The meaning of Eq. (3.11) is that the increment of the load is the product of a constant k multiplied the parameter t. The functions $c_{iT}(t)$ are arbitrary translations. The functions $F_{ij}(t)$ are the configurations of projected displacements ($i,j = 1,2$). Provided that the above conditions are satisfied in a metal subject to loading, the analyzed body remains elastic. In the moment that the functions that relate projected fringes and loads become function of the location X_{ij} and the parameter t, that is $F_{ij} = F(X_{ij}, t)$ or $c_{iT} = (X_{ij}, t)$, the metallic specimen enters the plastic state of deformation. Hence, nonlinear changes of fringe patterns of the projected displacement are an indication of the onset of plastic deformation of the observed metallic component. The onset of the plastic deformations is the first step in a process that ends with the actual failure of a metallic component. The failure can manifest itself in two ways, plastic collapse of the part where the part becomes a mechanism in motion under the collapse load. The other possibility is that the onset of plasticity is followed by actual fracture of the component.

3.3 Scale Dependence of the Experimental Observations of the Transitions to Plasticity and Fracture

Before dealing with the subject matter of the paper, it is necessary to make some consideration concerning the scale dependence of the observed experimental information and resolution dependence. One should understand that described phenomena, plasticity and fracture, depend on the observation scales. Getting quantitative information depends on the selection of correct kinematics and dynamic variables. Transition from linearized strain and stress tensors to the nonlinear ones poses important and significant differences in data analysis. Also spatial resolution plays a very important role: if the spatial resolution or displacement resolutions are both low or one of them is low, only average values will be detected.

The representative volume element (RVE) is a link between the discontinuous nature of materials and Continuum Mechanics. Materials are defined by their mechanical properties. These properties represent in Continuum Mechanics averaged values at certain subscale. The averages are computed at an area or volume with a given shape that for convenience in 2D is a square and in 3D a cube. Conditions that correspond to a given RVE are selected on the basis of the Hill-Mandel [12, 13], homogenization principle. It states that for given σ_{rve} and ε_{rve}, stresses and strains of the RVE, the virtual work in the macroscale equals the virtual work in the subscale. This principle is of great significance. The failure of fulfilling it results in important errors in the values of computed quantities as pointed out in [14–19].

The measurement of local kinematic and dynamic variables for a given RVE with large deformations and rotations requires removing limitations due to linearized kinematics and dynamics variables. Displacement fields cannot be described in a

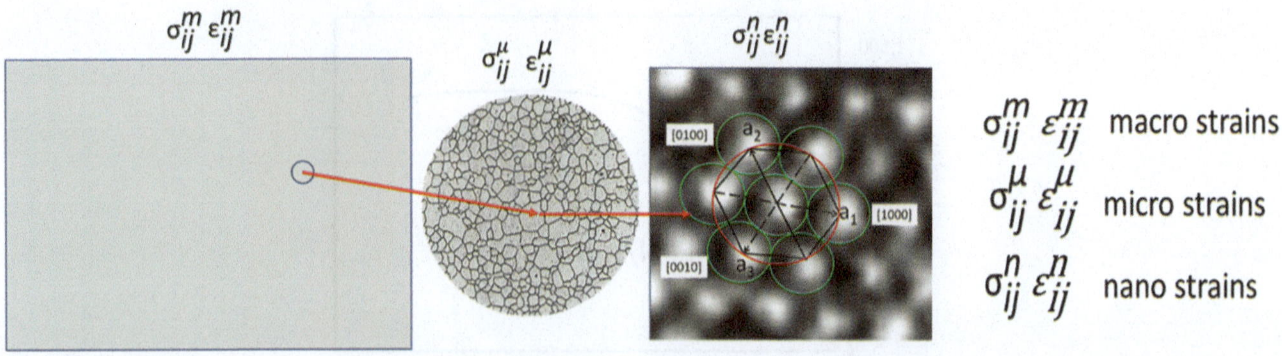

Fig. 3.3 Illustration of three scales of experimental observation of images captured in Experimental Mechanics

unique geometric reference system. One has to select a given RVE scale and the corresponding coordinate system to define local variables, choosing either a Lagrangian or a Eulerian representation. To discuss this aspect, we recall Fig. 3.3 from [20]. Figure 3.3 introduces graphically the relationship of measurement of mechanical properties at three different scales. In applications, many different scales can be introduced, each one will provide different aspects of the kinematics and dynamics of the observed materials. The observed fields are scale dependent and also depend on the spatial resolution that can be achieved.

The same field observed with different spatial resolution will provide different results depending on the gradients of selected variables in the field of interest. At a given scale, the behavior of a material may appear to be in the field of quasi-elastic behavior, while at a different scale one can observe local transition to plasticity or transition of plasticity to fracture. The following relationships between scales should be valid.

$$\sigma_{ij}^{m}(x) =< \sigma_{ij}^{\mu}(x) >= \frac{1}{V_R}\iint\limits_{V_R} \sigma_{ij}^{\mu}(x_R) \tag{3.12}$$

$$\varepsilon_{ij}^{m}(x) =< \varepsilon_{ij}^{\mu}(x) >= \frac{1}{V_R}\iint\limits_{V_R} \varepsilon_{ij}^{\mu}(x_R) \tag{3.13}$$

The meaning of Eqs. (3.12 and 3.13) is that the field average at a lower scale should give the values of the stresses and strains at the corresponding points of the upper scale.

Another requirement needs to be satisfied. It is necessary to adopt a stress tensor and a strain tensor compatible with each other in the Mandel-Hill sense. In Experimental Mechanics, since images are obtained in the deformed state, it is more convenient to work with the Eulerian description. In this case, the selected strain tensor should be compatible with the Eulerian description, and when strain tensor is selected, the stress tensor should be compatible with the strain tensor in the Hill-Mandel sense.

3.4 Transitions to Plasticity

In this section and following sections, we are dealing with space-temporal transitions of tensile specimens from elasticity to plasticity and the transition of plasticity to fracture. All these transitions are affected by the applied strain-rates to the specimen as well as by the interactions between testing machine and specimen. The effect of these two factors are outside the scope of this study. Examples will be analyzed that have been subjected to different strain rates chosen for research reasons other than these two mentioned variables. Indeed, these two factors may have influenced the analyzed images, but their effects are not discussed in the present work.

In Sect. 3.2, it has been established that the elastic behavior of a metal is characterized by an affine transformation making the distribution of displacements proportional to applied load. From the point of view of the kinematic of the continuum, variables are linearized expressions of strains, the simplified expression of the rigid body rotations, and the Hooke's law of that relates strains to stresses. The actual stress tensor utilized is the Cauchy stress tensor that is compatible with the Eulerian

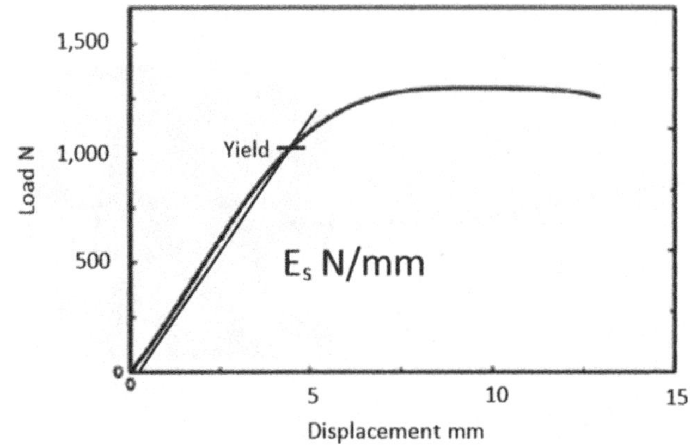

Fig. 3.4 Load vs. displacement plot for a tensile specimen

description of the continuum [20]. However, in the linearized Continuum Mechanics, the distinction between the Lagrangian and Eulerian description is dropped. Hence, in dealing with plasticity, one must be careful in the selection of variables to discuss the transition to plasticity. The transition implies a change in the description of the continuum and consequently of the variables that define a material between the two different behaviors. This is an important point to be considered if one wants to define the transition of state using displacement patterns.

A representation of force vs. displacement, Fig. 3.4, provides the classical method to detect transition from the elastic regime to the plastic regime. With a certain precision a line with the slope E_s (N/mm) is drawn, the elastic regimen is established and separated from the plastic regimen that at each point can be characterized by the local slopes. As explained in Sect. 3.2, for a given displacement δ (mm), the corresponding force is

$$P(N) = E_s \frac{\mathrm{N}}{\mathrm{mm}} \times \delta\mathrm{mm} \tag{3.14}$$

In Eq. (3.14), E_s is the slope of the plot. A similar criterion can be applied for a 2D displacement distribution utilizing properties of the isothetic lines (moiré fringes). In [7], it was introduced the family of fringes of equal projected displacements calling them isotachis fringes. Also, families of the derivatives of the isotachis fringes were introduced calling them isotachis of the partial derivatives.

In this paper, we will call these fringes iso-derivatives fringes. Calling U(**x**,P) and V(**x**,P) the projected displacements with respect to the coordinates, the symbol bold **x** represents the vector $\mathbf{x} = \mathbf{x}\mathbf{i} + \mathbf{y}\mathbf{j}$. The notations U(**x**,P) and V(**x**,P,) will be utilized to indicate families of fringes, u(**x**,P) and v(**x**,P) for individual values. As the applied load or displacement on a body is increased, the displacement field is modified. This change results in a change of the spatial velocity of the displacement field at a given point. In the linear case, for small deformation and rotation theory, these changes are components of the strain tensor. In the general case, these derivatives are no longer strains, but we call them spatial velocities or changes of the projected displacements per unit of length. We introduce the notation for spatial derivatives

$$\frac{\partial u\,(\mathbf{x},\mathrm{P})}{\partial x} = \varepsilon_{\mathrm{u}}\,(\mathbf{x},\mathrm{P}) \tag{3.15}$$

$$\frac{\partial \mathrm{v}\,(\mathbf{x},\mathrm{P})}{\partial y} = \varepsilon_{\mathrm{v}}\,(\mathbf{x},\mathrm{P}) \tag{3.16}$$

Hence, it is important to make a distinction between iso-derivatives, or loci of the points with the same spatial velocity and iso-strain lines, since only for small rotations and small deformations the strain components are linear functions of the derivatives of the displacements. The moiré method provides the means to generate iso-derivative fringes, and since the common way in moiré method is to project displacements, there will be two families of iso-derivative fringes. Iso-derivatives fringes are important tools to define the transition from elasticity to plasticity in 2D or 3D.

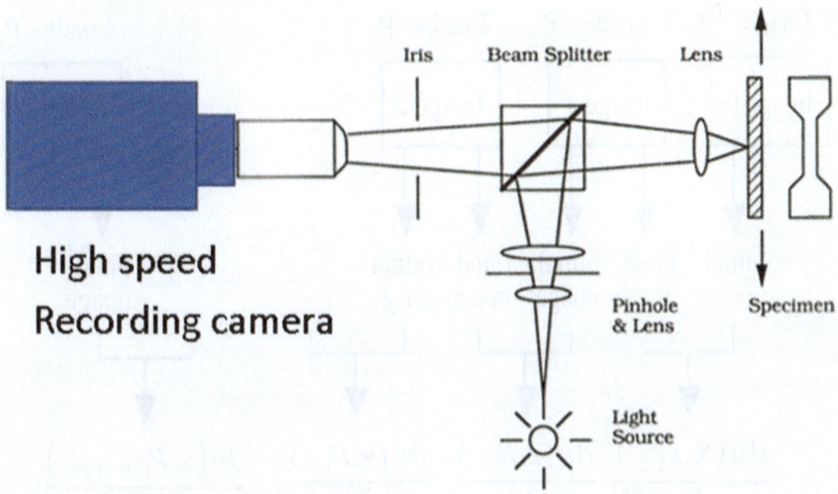

Fig. 3.5 Optical setup to obtain patterns of the iso-derivative fringes

The experimental determination of the iso-derivative fringes can be achieved by using different optical setups. The most general setup requires high-speed cameras for recording images as load or displacements are applied to the observed specimen. Also, it is necessary to introduce software that provides the derivatives of the displacements with respect to the coordinates x–y, (**x**), for increasing loads or applied displacements, for example, tagging on the specimen an orthogonal system of carrier lines. Then, one can resolve the carrier gratings utilizing, for example, the optical set up shown in Fig. 3.5 and non-coherent illumination. With this system, it is possible to resolve the printed grating at different scales by changing the magnification of the system, and by filtering it is possible to get the components ε_u (**x**, P) and ε_v (**x**, P). The necessary computations can be done by utilizing digital moiré [21]. The setup shown in Fig. 3.5 can be used to determine iso-derivatives fringes in the microscopic range.

An alternative way to determine the iso-derivative fringes is to use coherent illumination and two or four beams speckle interferometry setups. First, it is necessary to remember that, as shown in [21], speckle patterns provide isothetic lines as the moiré method does but with limitations arising from decorrelation and noise content.

Examples of the determination of the iso-derivative fringes utilizing speckle interferometry are given in [2–5] and [8–11]; in these references, tensile specimens were studied. In [4], two interferometers were utilized, one with the sensitivity $\mathbf{S}_1$ in the longitudinal direction of the specimen to make measurements in one face of the specimen. The other interferometer with the sensitivity vector $\mathbf{S}_2$ transversal to the specimen to make measurements in the other face of the specimen. Both interferometers record images by means of a high-speed camera.

The data processing is done so that the obtained patterns represent finite differences of the iso-derivative fringes. In [10, 11], only longitudinal displacements are recorded by a CCD camera, only one of the two families is determined at certain points the stress-strain curve. Figure 3.6 illustrates the schematic process utilized in [4] to obtain the finite differences of u(**x**,P), and a similar process can be applied to obtain v(**x**,P),

$$\frac{\Delta \mathrm{u}\,(\mathbf{x}, \mathrm{P})}{\Delta x} \approx \varepsilon_{\mathrm{u}}\,(\mathbf{x}, \mathrm{P}) \tag{3.17}$$

$$\frac{\Delta \mathrm{v}\,(\mathbf{x}, \mathrm{P})}{\Delta y} \approx \varepsilon_{\mathrm{v}}\,(\mathbf{x}, \mathrm{P}) \tag{3.18}$$

In [4], the utilized specimen is a thin foil plate 100 mm long and 0.4 mm thick, a thin specimen. To define a critical area where the transitions to plasticity and fracture are localized, one side of the specimen is made curved. The upper width of the specimen is $w_o = 20$ mm and the middle section $w_m = 15$ mm. This design causes the tensile specimen to be subjected to tension and also to bending (see Fig. 3.7). This fact is important because the obtained results depend on the geometrical configuration and boundary conditions. Specklegrams are acquired by a movie camera at the speed of 30 frames/s, and the specimen is pulled at the speed of 4 μm/s. The acquisition software is designed in such a way that according to the scheme of

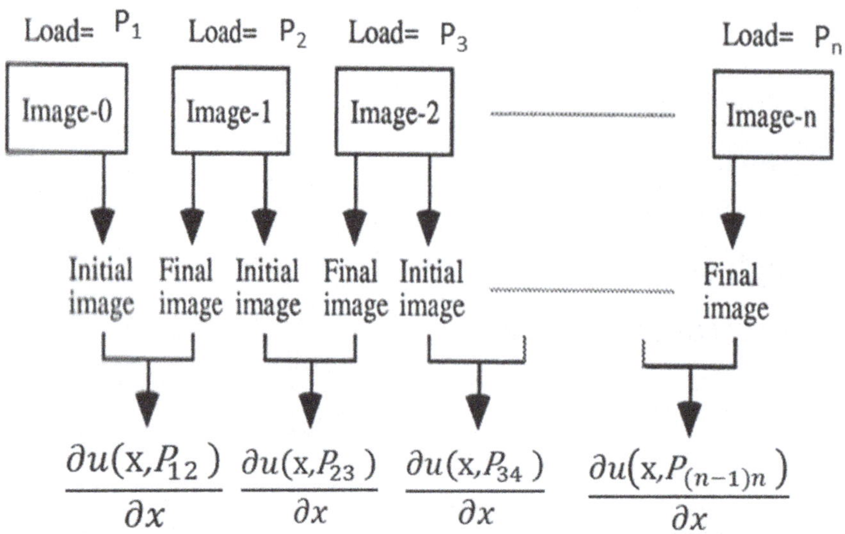

Fig. 3.6 Schematic representation of the determination process of iso-derivatives fringes at different load levels for u(**x**,P). A similar scheme is utilized for displacements v(**x**,P)

Fig. 3.7 Schematic of a tensile specimen subject to axial force and to bending stresses

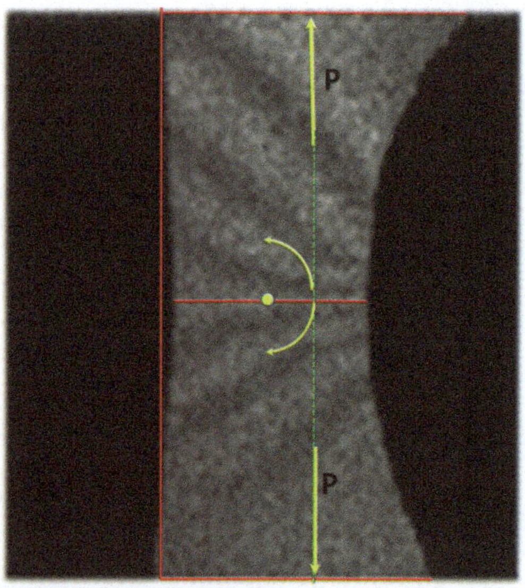

Fig. 3.6, the image corresponding to a given frame and a following frame are subtracted: Eqs. (3.17) and (3.18) are obtained, thus generating iso-derivative fringes. The specklegrams contrast depends on the correlation between the two interacting speckle patterns. Acquisition times and loading times are such that good contrast fringes are obtained.

For example, in the case of the V displacements, isothetic lines in the direction of the vertical axis of the specimen are functions V(y,P,t) where t is the time corresponding to a frame number of the recorded images, by the chain rule of differentiation,

$$\frac{\partial \mathrm{v}\,(\mathrm{y},\mathrm{P})}{\partial t} = \frac{\partial \mathrm{v}}{\partial y}\frac{\partial y}{\partial t} + \frac{\partial y}{\partial P}\frac{\partial P}{\partial t} \tag{3.19}$$

Equation (3.19) indicates that if we have a film recording of displacements and times for these displacements, and at the same time recordings of the loads corresponding to the film frames, we can get also the time derivatives of the projected velocities of deformation. A similar equation can be written for the U(x,P) component.

3.5 Signals Recorded by the Camera

At this point it is necessary to point out that the signals recorded by the camera in the optical system are not only space dependent as it occurs in static cases but are time dependent. In the process of loading a tensile specimen, one has to compare the velocity of the applied load or displacement and the process of wave propagation in the specimen. The speed of wave propagation in the specimen in the longitudinal direction depends on the particular region of the load vs. displacement plot (Fig. 3.4). In Fig. 3.8, the iso-derivative fringes of the tensile specimen studied in [4] are displayed. As the frames number change, the loci of the iso-derivatives change. The iso-derivative fringes of ε_v (**x**, P) have the same trend as the displacements V(**x**,P) due to the axial displacement of the specimen, and the iso-derivatives fringes of ε_u (**x**, P) are the result of the Poisson's effect and have the same trend as U(**x**,P). It is possible to see, label (a) of lower row, that due to the bending effect, the iso-derivative fringes of the U(**x**,P) are not vertical as they ought to be in a tensile specimen but are slightly curved loci due to the bending.

In Fig. 3.9, the plot load P vs. frame number of the recorded images of the iso-derivative fringes is shown, through the added labels one can see the images in Fig. 3.8 corresponding to the selected frames. Since the frame rate is given, the plot of Fig. 3.8 is also a plot of P(**x**,t).

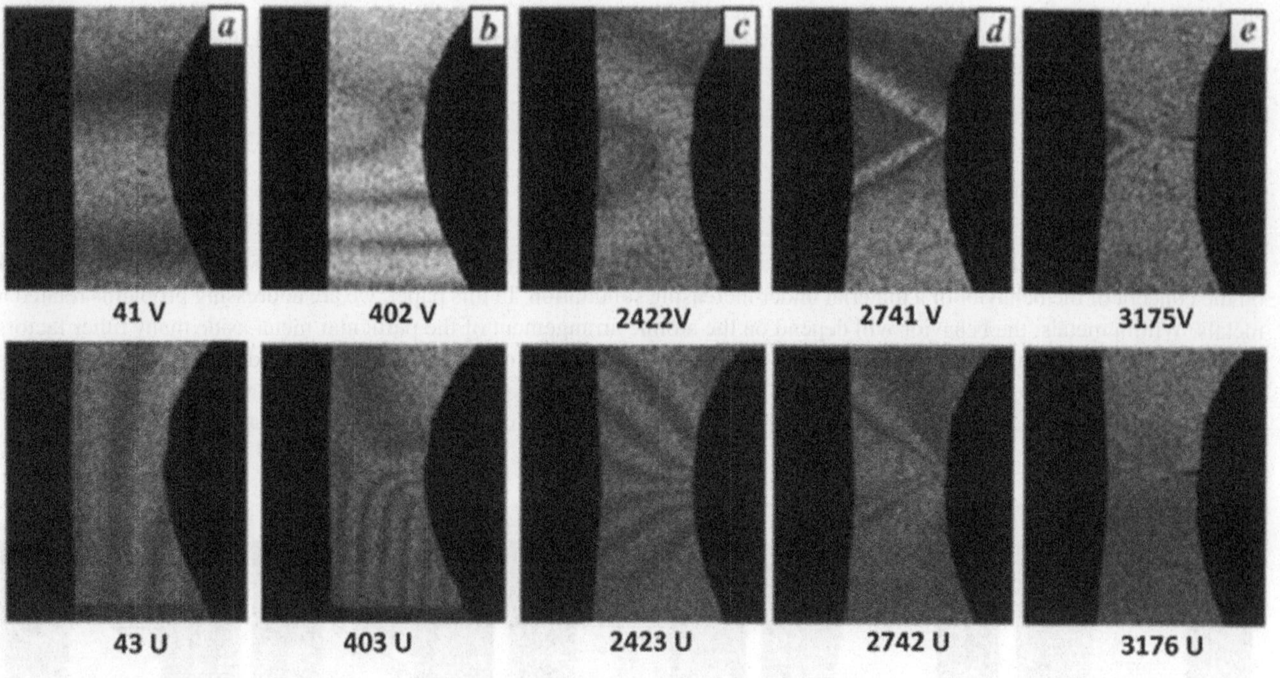

Fig. 3.8 Display of the finite differences iso-derivatives ε_v (**x**, P) (upper row) and ε_u (**x**, P) (lower row). Labels indicate the displayed frames corresponding to Fig. 3.9

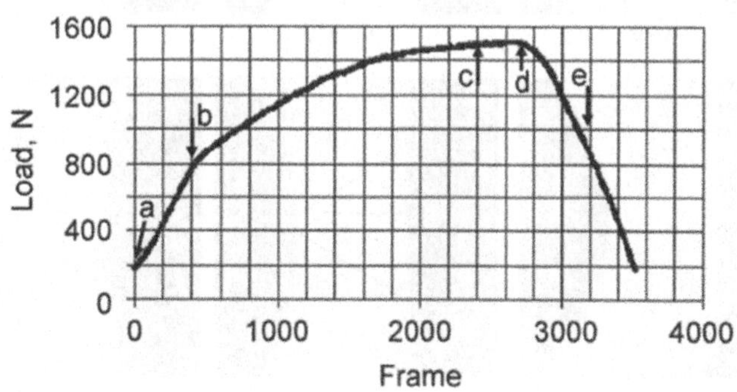

Fig. 3.9 Plot of the relationship between acquired frames and applied load as indicated by the testing machine. Labels a, b, c ... correlate this curve with the iso-derivatives fringe families of Fig. 3.8

Observing Fig. 3.8(b) upper row and Fig. 3.9, the image corresponds to the onset of plasticity. At the edge of the specimen, $\varepsilon_V(\mathbf{x}, P) = 0$. The linear relation between iso-derivatives fringes and loads ends. Also, lower row, Fig. 3.8 label (b), $\varepsilon_u(\mathbf{x}, P) = 0$ indicates transition to plasticity. Hence, the transition to plasticity indicated by label (b) in Fig. 3.8 is well detected by the two families of iso-derivatives fringes.

3.6 Transition from Onset of Plasticity to Fracture

In Sect. 3.5, it is concluded that the iso-derivative lines can be utilized to detect the onset of plasticity recording either $\varepsilon_u(\mathbf{x}, P)$ or $\varepsilon_V(\mathbf{x}, P)$. The next question to analyze is the detection of the onset of fracture. This can be done analyzing the frames of Fig. 3.8, label (c) iso-derivative fringes before maximum load, label (d) iso-derivative fringes at the maximum load, and label (e) the iso-derivative fringes in the process of post fracture.

One clue in this transition can be extracted from the slopes of the iso-derivative fringes (Fig. 3.10). At label (c) near the maximum load, the slopes of the iso-derivative fringes display asymptotic values that are the same for both families of iso-derivative fringes. The iso-derivative fringes of the U(**x**,P) have increased in number with respect to the iso-derivative fringes of V(**x**,P). At label (d), the maximum of the tensile load takes place. The sudden change from (c) to (d) indicates that an abrupt change in the specimen geometry has taken place.

The next step is to follow the transition from yielding to fracture. In order to achieve this goal, it is necessary to review some concepts.

3.7 Instability of the Plastic Flow

The analysis of the transition from the onset of plasticity to actual fracture is an extremely complex subject because it depends on the concept of the behavior of a material under increasing solicitation. In this paper, we are addressing problems related to metals. Within metals, the behavior will depend on the atomic arrangement of the particular metal aside many other factors. We are further narrowing our field to tensile specimens that are utilized in standards of testing to define material properties. We can analyze the concept of hardening, i.e., if in Fig. 3.4 in place of plotting P vs. δ, one plots a version of the stress σ vs. a corresponding expression of strain ε, one obtains a definition of hardening or equivalent to a local modulus,

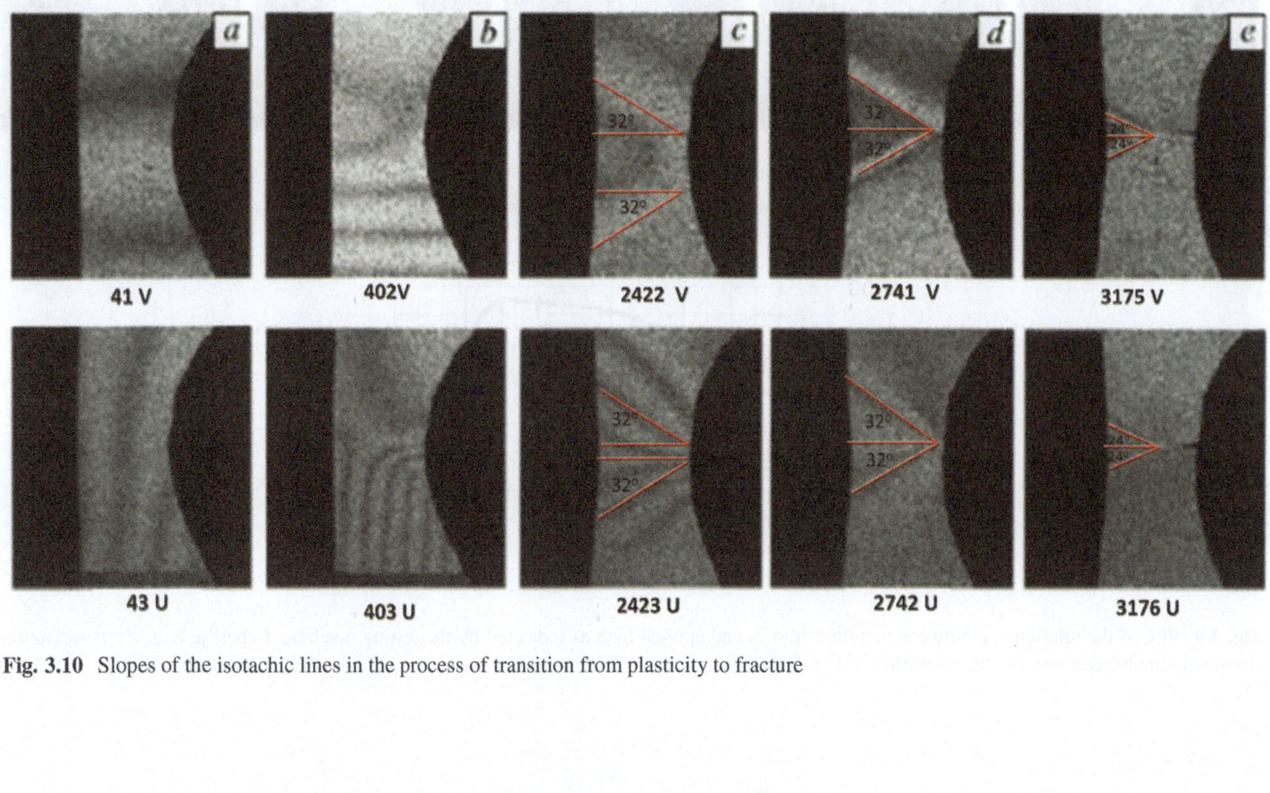

Fig. 3.10 Slopes of the isotachic lines in the process of transition from plasticity to fracture

$$h_d = \frac{d\sigma}{d\varepsilon} \tag{3.20}$$

If Eq. (3.20) is positive, it indicates an increase of the load carrying capability of the specimen.

The hardening process may be the result of many different variables, but the net effect is that a positive $\Delta\sigma$ for an increment of a strain $\Delta\varepsilon$ indicates that the material can take an increasing load. The condition for maximum load that the specimen shown in Fig. 3.7 corresponds in Fig. 3.8 to

$$\frac{d\sigma}{d\varepsilon} = 0 \tag{3.21}$$

Equation (3.21) is an indication that the specimen has reached the maximum load and the subsequent negative values of the slope shown in Fig. 3.8, (e) indicates a softening. The softening can be the result of material changes or geometrical changes or a result of both effects. Considering that the carrying load capability is given by $P = \sigma A$, it follows:

$$\frac{d\sigma}{d\varepsilon} = \frac{1}{A}\frac{dP}{d\varepsilon} \tag{3.22}$$

In the case of tensile specimens, structural instability manifests through a process denoted in the literature with the generic name of necking, a sizable change of the cross section of the specimen that follows the yielding of the material, a reduction of the cross-sectional area of the specimen. The actual shape of $\sigma = f(\varepsilon)$ depends on many different variables. In the case of very ductile specimens, maximum load is followed by an immediate softening of the material, a negative slope. The necking process depends on the geometry of the specimen, loading process, and molecular structure of the metal among the main variables. Since our objective is the interpretation of the patterns of iso-derivative fringes, let us analyze patterns that can provide further clues on this process.

Changes of geometry caused by plasticity are directly connected to dislocation dynamics and hence on molecular structure of considered metals. That is, in the last instance, this process involves the active systems of slip lines that can produce geometry changes compatible with a given molecular structure adapting to the applied solicitation and generating the strain hardening required to balance the applied load until this process is no longer viable.

In order to link images captured by a high-speed camera to mechanisms that lead to the onset of fracture utilizing available experimental observations, it is interesting to start with an interpretation of a sequence of recorded images. This is done in Fig. 3.11 with a scheme that graphically relates events at the molecular level and the recorded behavior of a tensile specimen.

The patterns shown in Fig. 3.11 [2] correspond to a tensile specimen made of an aluminum alloy A5052H112. The specimen gage length is $L_o = 150$ mm, width is $w_o = 25$ mm, thickness is $t_o = 2$ mm, ratio width to thickness is $r_{wt} = w_o/t_o = 12.5$, and pulling speed of the machine is $v_{pm} = 5.83$ μm/s.

Figure 3.11 represents actual observations of the tensile specimen together with a schematic representation of the process leading to the fracture of the specimen after it has yielded. The tensile specimen is fixed at one end and the testing machine displaces the other end at the mentioned uniform crossbeam speed. At the beginning of the test, the pulled end generates elastic waves that are reflected at the fixed end, and the stresses of the specimen is cyclically increased. In the plot motion picture frames vs. applied load of Fig. 3.11, the loading process corresponds to points in the interval labelled (a)–(b), elastic range. At point (b), the onset of plasticity takes place, and a wide band appears at the loading end and propagates towards the fixed end; this band is called in the literature Lüders band. The front end of the band makes angles typically from 32° to ±7° (see Fig. 3.10). The band represents the localized plastic deformation described at the beginning of the section. The band propagates with a speed that is related to the speed of motion of the loading machine. In Fig. 3.11, it is assumed that due to symmetry conditions, if the specimen is homogeneous, propagation of the band decelerates to a stop at the center of the specimen where remains stationary and the process of fracture unfolds. The maximum load has been reached, point (d). In the neighborhood of (d), crack propagation begins. If iso-derivative fringes are recorded (Fig. 3.11), the process of fracture causes an unloading of the specimen, and the iso-derivative fringes re-appear in the specimen. The last frame reproduced in Fig.3.11 shows the broken specimen beyond point (e), and the fracture pattern makes an angle of 24° with the horizontal direction.

Figure 3.11 contains a symbolic representation of dislocations motion in the metal of the specimen. A discontinuity appears at the external edge of the specimen. The edge dislocation represent the presence of defects in the structure. At (a), when the elastic loading starts dislocations although present in the specimen are not activated. At (b), the dislocations begin their motion under the applied shear forces. Since the specimen is polycrystalline, these events take place in grains whose orientation is favorable to the motions of dislocations (frames (b)–(c)). At a given point in the specimen loading, enough dislocations have joined together to reach the free surface of the specimen and merge with the roughness of the surface. This

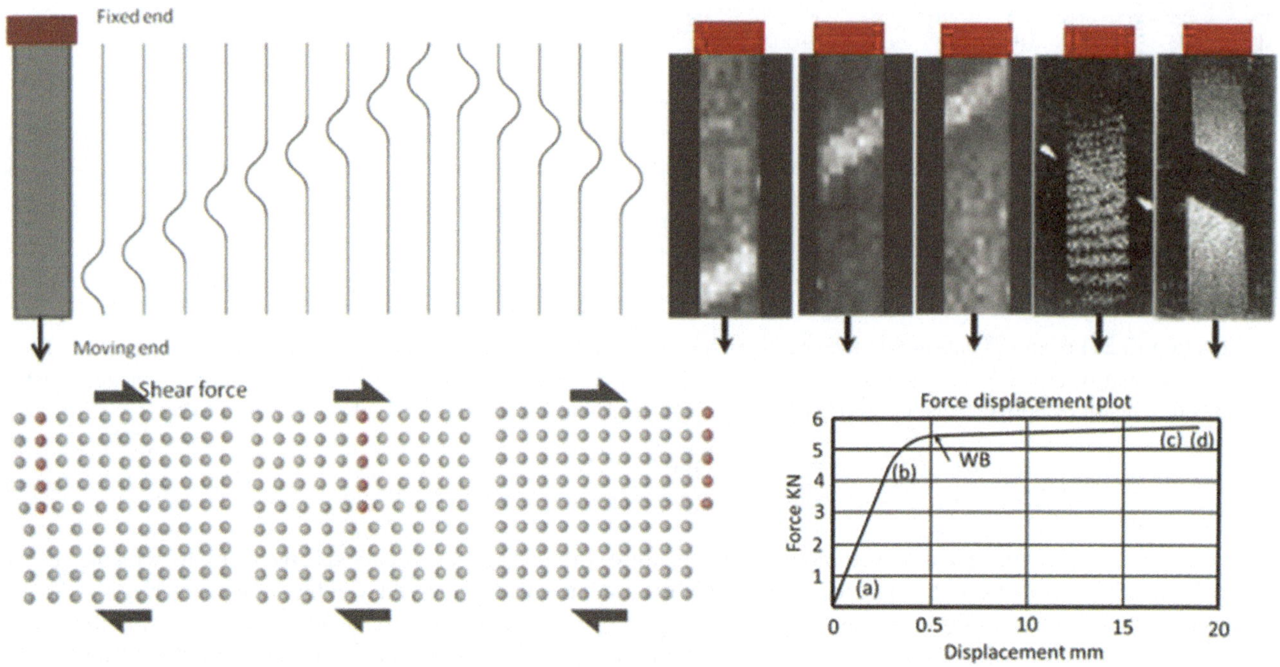

Fig. 3.11 Sequence of events leading to the ductile fracture of a metallic specimen

event is schematically represented by a discontinuity in the surface of the specimen (Fig. 3.11). The sequence illustrated in Fig. 3.11 is important to help in understanding the relationship between related events. For the example of Fig. 3.11, the transition from yielding to fracture can be separated into two processes: the propagation of the structural instability from the place of initiation of a plastic wave to the final location in the specimen where the actual onset of fracture takes place. The processes taking place between the initiation of the plastic wave propagation and the beginning of fracture are extremely complex, and there is a large inventory of possible outcomes.

Before one can proceed in the analysis of the previously described processes leading to the fracture of a specimen, it is necessary to understand the link of captured images and physical events that are connected with these images.

3.8 Analysis of the Process of Propagation of the Plastic Instability

Figure 3.12 illustrates the effect of the onset of the structural stability in a specimen of rectangular cross section. Both w_o and t_o are reduced and the sides of the rectangular section become curved arcs. The changes in dimensions of w_o and t_o depend on the magnitude of the axial force *P*. The magnitude of these changes depend on many different variables, but this is the general trend of the signal that propagates as a wide band along the specimen as it is shown in Fig. 3.11. The recorded shape of the band depends on the space and time resolution of the optical system utilized to track the propagation of the plastic instability.

To relate the dimensions of the wide band to actual measurements, data taken from [10] are plotted in Fig. 3.13. These data correspond to the maximum load of the specimen of Fig. 3.14 when the plastic wave propagation stops, and the fracture process starts.

Data shown in Fig. 3.13 correspond to the specimen called No. 2 in Ref. [10], Fig. 3.14, with the following dimensions, thickness $t_o = 1$ mm, width $w_o = 3.6$ mm, and gage length $L_g = 19$ mm. The specimen is made of faced centered cubic (*f.c.c.*) austenitic 316L stainless steel. The analyzed image is a snapshot with exposure time $t_e = 20.36$ s, and the velocity of loading of the testing machine $v_{pm} = 0.1$ μm/s. The data were obtained analyzing the iso-derivative ε_v (**x**, P) pattern of Fig. 3.14 that corresponds to the point of maximum load in the strain vs. load plot. Figure 3.13a shows the changes of w_o measured at $\Delta t = 20.36$ s. Figure 3.13b shows the plot of the ratio $r_w = w/w_o$.

The change in w_o at the center of the specimen is $\Delta w_o = \frac{0.29}{3.6} = 0.08$ or 8%. The spatial and temporal resolution satisfy the conditions required to capture the iso-derivatives that cannot be recorded in lower resolution images. Following the

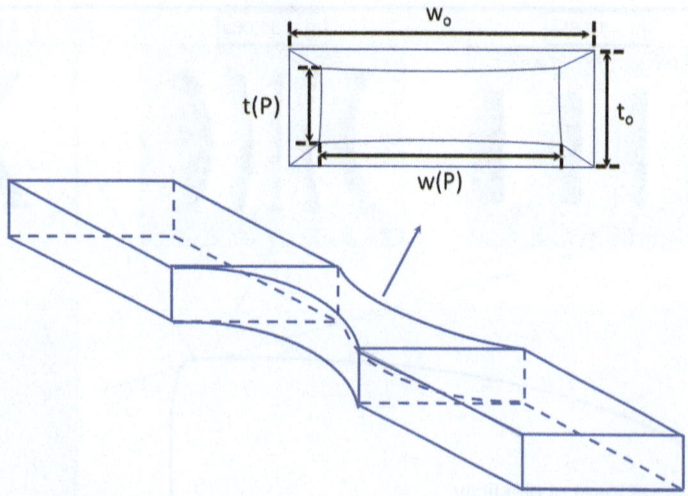

Fig. 3.12 Shape of a rectangular specimen when the plastic instability sets on, the rectangular cross-sectional dimensions w_o and t_o are reduced

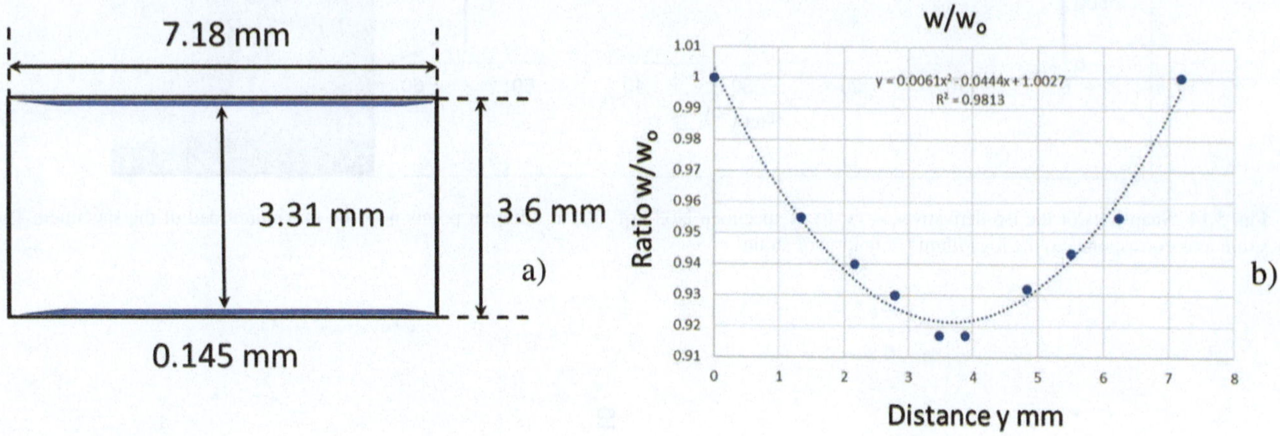

Fig. 3.13 Cross section in the axis of symmetry of specimen [11]

notation in preceding sections, the longitudinal axis of the specimen coincides with the y-axis of the Cartesian coordinates-axes and the corresponding displacements belong to the V(x,P) family of the iso-derivatives. The sensitivity $\mathbf{S}_1$ along the specimen axis has a modulus $|\mathbf{S}_1| = 0.447\ \mu$m.

Figure 3.14 shows the iso-derivatives fringe patterns in the segments of the plot force F vs. the strain around the maximum load and also includes a picture of the broken specimen.

The snapshot of Fig. 3.15a corresponds to an area where the wave front has already passed through and shows iso-derivatives corresponding to the pattern "V" of Fig. 3.1. Since the rate of change is decreasing, spatial derivatives correspond to the condition $\frac{d\sigma}{d\varepsilon} \to 0$.

The length of the neck region at maximum load from Fig. 3.13b is $L_n = 7.18$ mm, and the ratio between L_n and w_o is $r_{lw} = 7.18/3.6 = 1.9$. This result indicates that within the dispersion of the experimental results, this ratio is approximately equal to 2. This value is consistent with the Saint-Venant's principle of decay of solutions of four order differential equations [22]. This approximation is confirmed by the results plotted in Fig. 3.16 of the final shape of the neck after the fracture of the specimen, $L_n = 7.7$ and $r_{lw} = 7.7/3.6 = 2.14$.

From the above-presented data analysis, the following overall view of the propagation of the plastic instability emerges in a material that is homogeneous and is loaded under very stable conditions. At the onset of plasticity, Fig. 3.14, the shape of the specimen changes to a configuration similar to the configuration that is outlined in Fig. 3.12 and that in the literature is referred as diffuse necking. As the load is increased, the geometrical configuration is basically preserved but the magnitude of w(P) and t(P) change until the maximum load is reached. At this point, the wave propagation stops and the process of fracture begins.

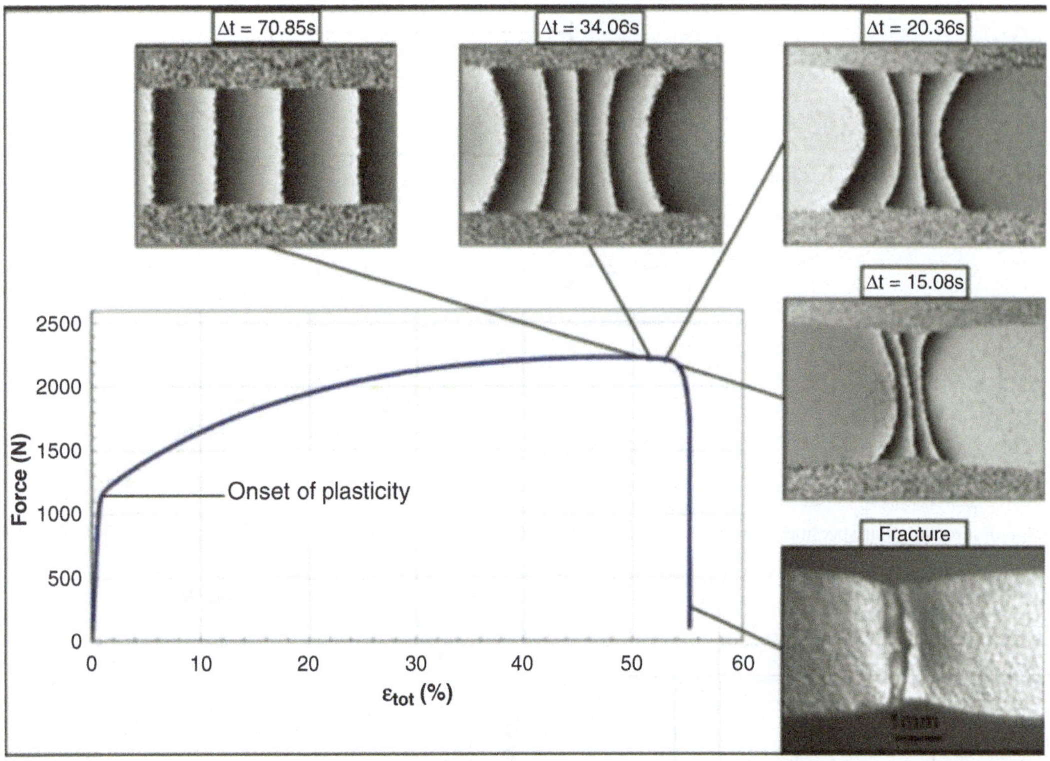

Fig. 3.14 Snapshots of the iso-derivatives ε_V (**x**, P) of specimen No. 2 of [10] at different points near the maximum load of the specimen. The strain axis corresponds to the logarithmic definition of strain

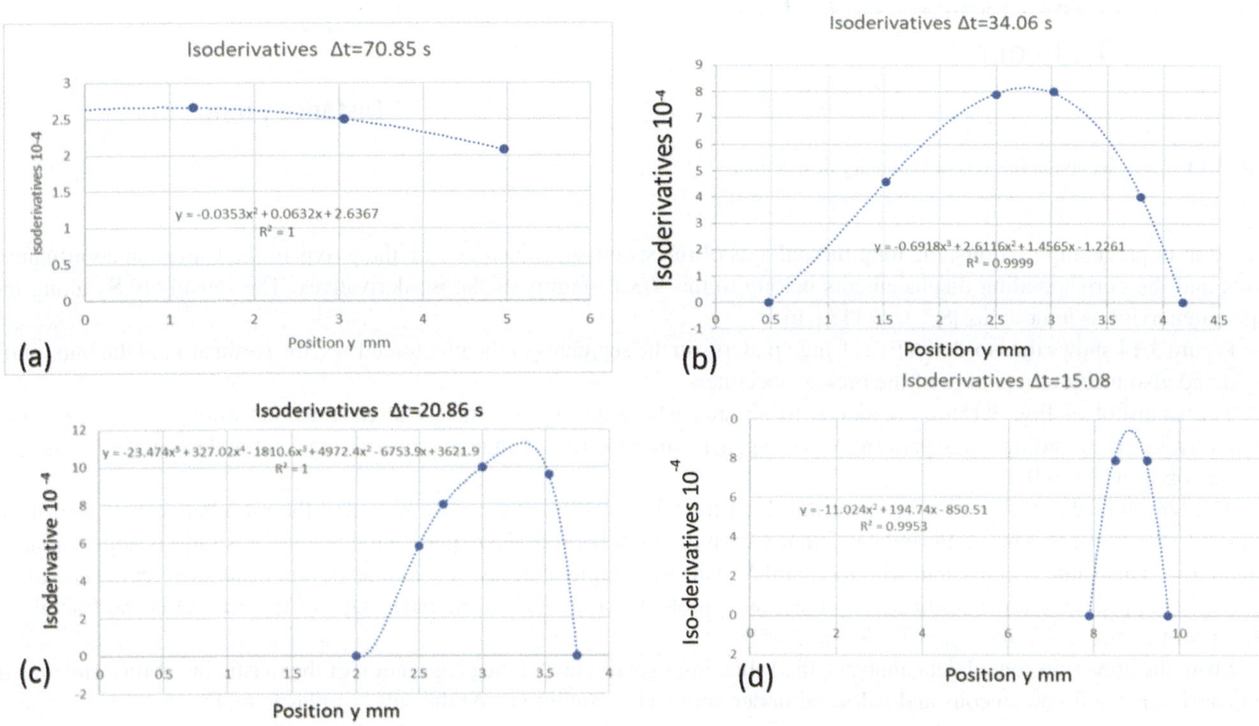

Fig. 3.15 Plot of the iso-derivatives in the specimen no. 2 [10], corresponding to the snapshots of Fig. 3.14

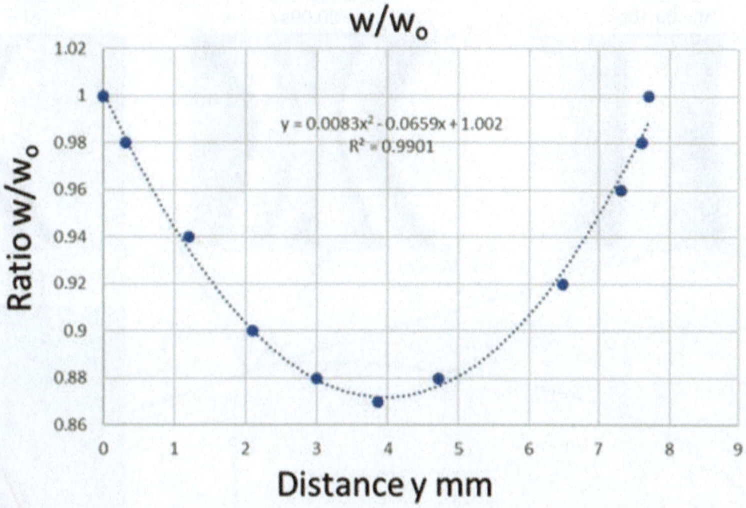

Fig. 3.16 Plot of the ratio w/w_0 corresponding to the top view of the shape of the specimen after fracture

Table 3.1 Data corresponding to fracture process

Δt s	L_{iso} mm	$\varepsilon_1^E \times 10^{-4}$
34.06	3.70	8.00
20.86	1.75	11
15.68	0.75	9.4

3.9 Fracture Process

In Sect. 3.8, the propagation of the structural instability from the point of view of geometrical changes is analyzed. In what follows, the transition from propagation to fracture will be considered. First, it is necessary to look at some notation clarification. The plots of Fig. 3.15 correspond to the center line of the specimen of Fig. 3.14, and the peak of the iso-derivatives are located very close to the axis of maximum contraction of the specimen (Fig. 3.13b). Since this center line is an axis of symmetry of the specimen, it coincides with the principal directions of the stresses and strains in this section. That is, the center line of the family V(**x**,P) coincides with the y-coordinate. Hence, we can write

$$\varepsilon_V \approx \varepsilon_1^E \tag{3.23}$$

Equation (3.23) indicates that the value of the iso-derivative in the finite differences sense is equal to the Eulerian principal direction along the axis of symmetry of the specimen. Equation (3.23) provides the principal iso-derivative along the centerline of the specimen and extends to most of the depth of the specimen. However, at the edges of the specimen, Eq. (3.23) is no longer valid and the plotted values correspond to derivatives including contribution of the rigid body rotations [20].

The data of Fig. 3.14 corresponding to the fracture process are shown in Table 3.1. The first column gives the recording times Δt, the second column gives the length of the segments L_{iso} where the iso-derivatives are different from zero and the third column gives the peak values of the iso-derivatives.

At $\Delta t = 34.06$, the maximum load has been reached. The region where the rate of change of the iso-derivatives is different from zero is given in the second column. In the following time intervals, the segments where the iso-derivatives are different from zero keep narrowing. This fact indicates that the local deformation increases at smaller segments in size until the fracture takes place.

It is interesting to see that the crack in this specimen is perpendicular to the axis of the specimen. That is in the language of Fracture Mechanics and in the approximation of the plane stress condition, the crack corresponds to a mode-1 fracture, although in reality the fracture surface is a 3D surface. Mode-1 fracture indicates that the axial force remains at the centroid of the specimen.

In the experimental observation of wide bands, in most of the cases the wide bands are inclined with respect to the center line. To analyze this case of wide band propagation, the specimen called specimen 1 of Fig. 3.17 [10] will be examined. In

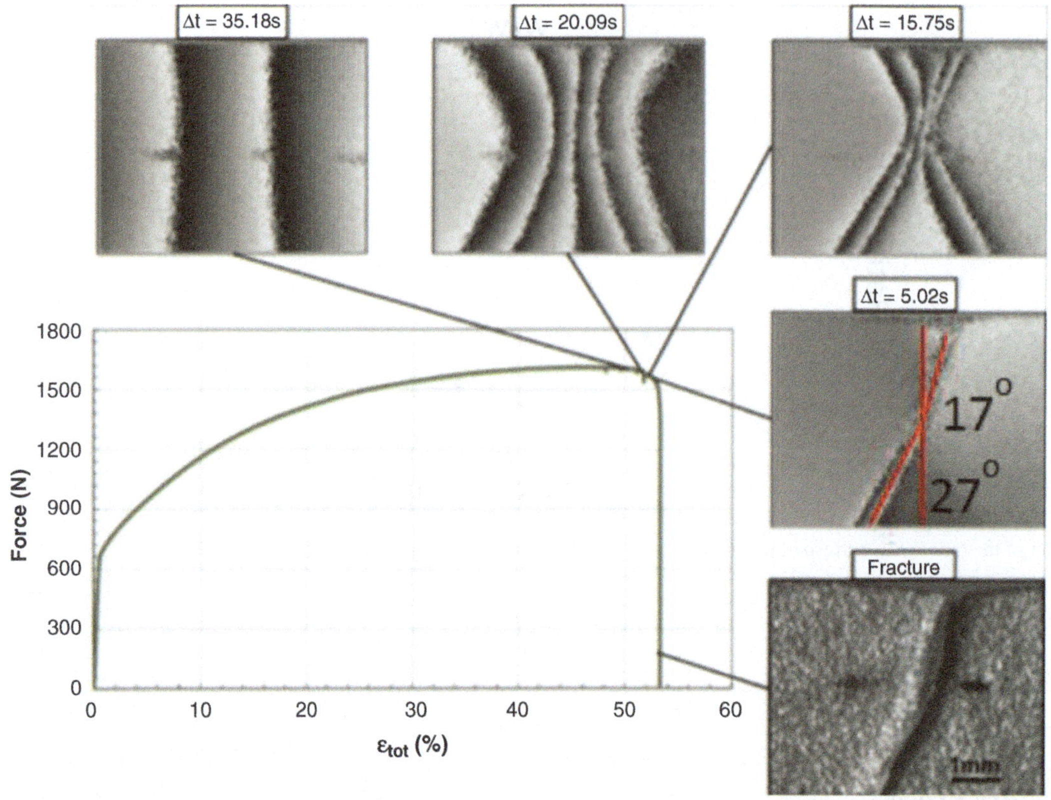

Fig. 3.17 Snapshots of the iso-derivatives ε_y of specimen No. 1 [10] at different points near the maximum load of the specimen. The strain axis corresponds to the logarithmic definition of strain

[11], a lot of additional information is provided for tests of these types of specimen that is very useful for completion of the subject matter under analysis.

The parameters characterizing this specimen are $L_o = 22$ mm, $w_o = 5$ mm, and $t_o = 0.5$ mm. The material of this specimen is the same material of specimen No. 1. Different speeds of the testing machine were used for different tests of this geometry; the optical sensitivity is $S_y = 0.447$ μm. A very interesting feature of the setup is the addition of acoustic emission (AE) equipment to capture the signals produced by the motion of dislocations and coordination with the optical signals. As it has been done with the other specimen, the iso-derivatives for the different time intervals Δt have been computed and are displayed in Fig. 3.18. For this specimen, it occurs $L_n = 1.93$, that is, close to 2 as it occurs for specimen No.1.

For $\Delta t = 35.18$ s, the iso-derivative shows a uniform rate of change of 20×10^{-4}, indicating that as it was determined for specimen No. 2, the wave front has passed through, and the iso-derivative pattern corresponds to the displacements "V" of Fig. 3.1.

For $\Delta t = 20.09$ (Fig. 3.18), the wave front stopped and the graphs represent the regions where the rate change of the iso-derivatives is concentrated around the middle of the neck formation. Graphs (a) and (b) give the upper edge and the middle section of the neck; while at the lower edge, the values of the iso-derivatives are smaller than those of the upper edge, the deformation at the upper edge is larger than at the lower edge.

For $\Delta t = 5.02$ s (Fig. 3.19), the maximum rate of change occurs in at the upper edge, and the slopes of the iso-derivatives of Fig. 3.17 are 27° and 17°, and the change of angle takes place at the middle of the section. The fracture crack reflects this change of direction. The type of fracture in the 2D plane stress condition indicates a mixed mode-1 and mode-2 fracture. This type of fracture indicates the presence of bending in the plane of the specimen increasing the strain in one edge and upper edge and reducing the stress in the lower edge.

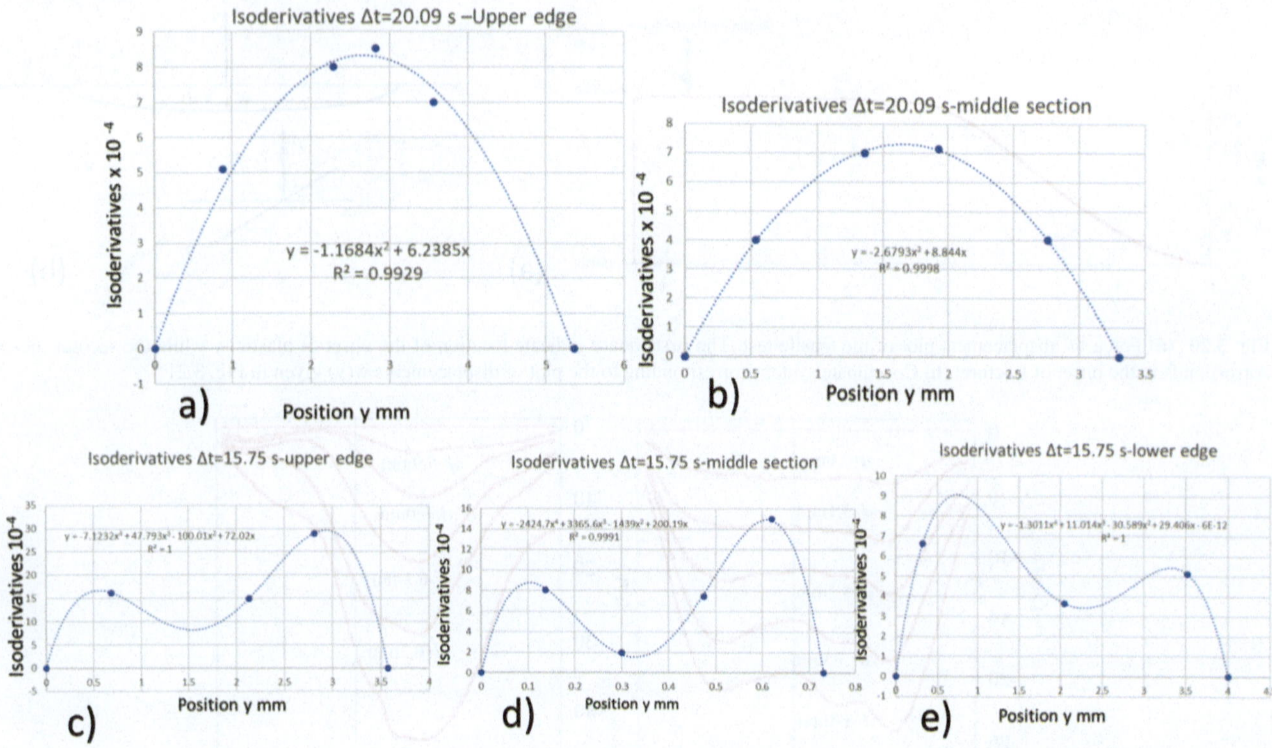

Fig. 3.18 Iso-derivatives plots from snapshots of the specimen of Fig. 3.17: (**a**), (**b**) $\Delta t = 20.09$ s; (**c**), (**d**), (**e**) $\Delta t = 15.5$ s

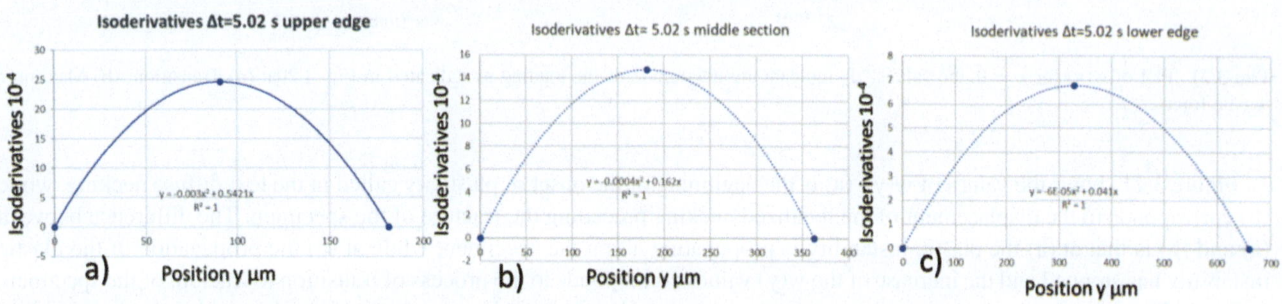

Fig. 3.19 Iso-derivatives plots from snapshots of the specimen of Fig. 3.17. $\Delta t = 5.02$ s, (**a**), (**b**), (**c**) $\Delta t = 15.5$ s

3.10 Changes in Depth Produced by the Onset of the Plastic Instability

In Fig. 3.12, it is shown that the narrowing of the cross section of a rectangular specimen in tension takes place also in the thickness of the specimen. Data on changes of thickness in rectangular cross-sectional specimens utilizing the moiré method and the fringe projection technique with a pitch of $p = 1$ mm are given in [23]. The tensile specimen parameters are $w_o = 20$ mm, $t_o = 0.6$ mm, and $L_g = 20$ mm, the speed of pulling is $v_{pm} = 8.33$ μ/s. The material of the specimen is brass that has also a f.c.c. structure.

This test differs from preceding examples in the sense that it is a static test; recordings are taken at increasing loadings of the tested specimen. While in the previously analyzed examples, the sensitivity vector corresponded to in-plane components of the displacement vector, in this example the sensitivity vector is in the direction perpendicular to the plane of the specimen. This means, using the literature conventional nomenclature for displacements perpendicular to a plane of reference, the displacements correspond to the W(**x**,P) family. The values of w(**x**,P) given are referred to an initial surface corresponding to the end of the elasticity range, in the beginning of the square region called in Fig. 3.20 transition region. The displayed displacements in Fig. 3.21 are total displacements with respect to the mentioned reference plane and not displacement derivatives as in the previously analyzed examples.

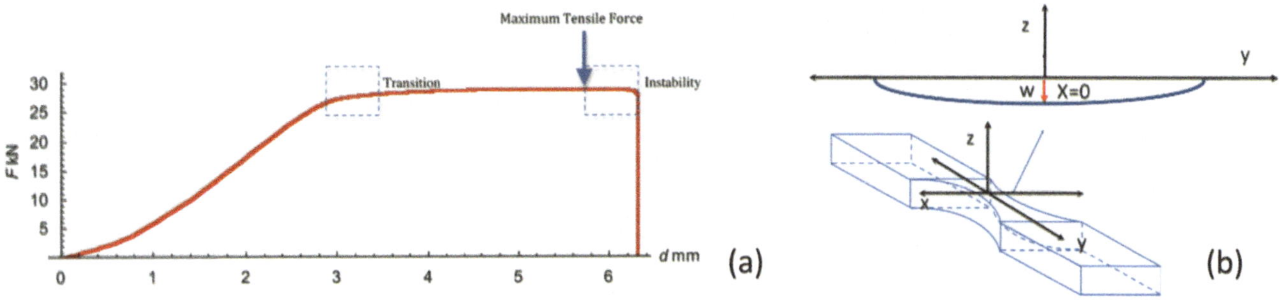

Fig. 3.20 (**a**) Force vs. displacement plot of the tensile test. The first square indicate location of the onset of plasticity while the second square corresponds to the onset of fracture; (**b**) Coordinate system corresponding to the plot of displacements w(y) given in Fig. 3.21

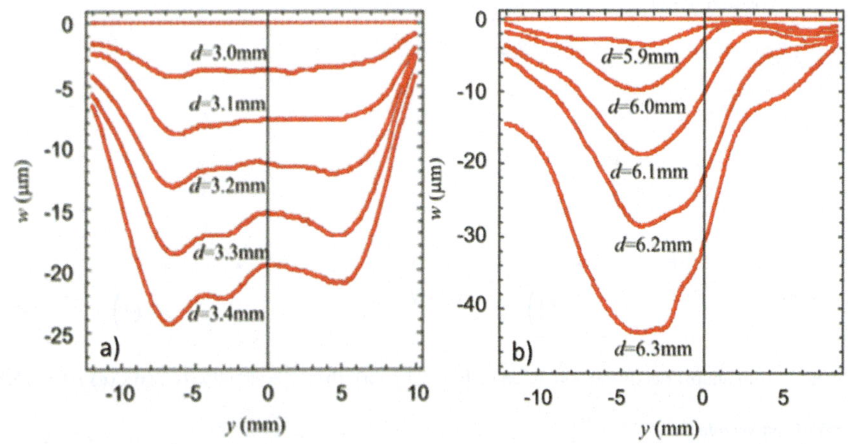

Fig. 3.21 Plot of w(y) for x = 0, the values "d" indicate the magnitude of the loading as indicated in Fig. 3.20a: (**a**) Transition; (**b**) Maximum tensile force

Figure 3.21 shows the values of w(y): (a) is the beginning of the onset of plasticity called in the text diffuse necking, while (b) corresponds to the displacement of the localized necking preceding the fracture of the specimen. The difference between (a) and (b) is that at (a) the plastic instability is propagating along the specimen, while at (b) the propagation of the plastic instability has stopped and the increase of the w(y) values corresponds to the process of transition to fracture of the specimen. In Fig. 3.21a, it is possible to see that, as the load is increased, the magnitude of w(y) is increasing as the plastic instability moves along the specimen. In Fig. 3.21b, the plastic instability is stationary but as the load is increased the values of w(y) increase. The values of w(y) correspond to the observed face of the specimen, assuming that the displacements of the back face are similar to those of the front face. The total maximum change of thickness of the specimen is $\Delta t_o = \frac{41.7 \times 2 \mu\text{m}}{600\ \mu\text{m}} = 0.14$ or 14%. This percentage change of thickness is very close to the change in width of the specimen of Fig. 3.14, 13% in the same region of the force vs. strain displacement plot.

From analysis of experimental results in Sects. 3.8 and 3.9, it can be concluded that the presence of signals corresponding to the plastic instability and detected with diverse techniques is correlated to the geometrical changes in the dimensions of the observed specimens called with the generic names of diffuse necking or fully developed necking. It also possible to conclude that the optically observed signals depend on the spatial and temporal resolutions utilized in the observation.

3.11 Propagation Velocity of the Plastic Wave Front

In the preceding section, we have analyzed the plastic instability from a geometrical point of view. Now we are going to deal with the fact that we are analyzing a wave propagation phenomenon from the onset of plasticity to the point where the transition to fracture takes place.

Following the same approach of the preceding sections, we will start with experimental data contained in [24] and graphically represented in Fig. 3.22.

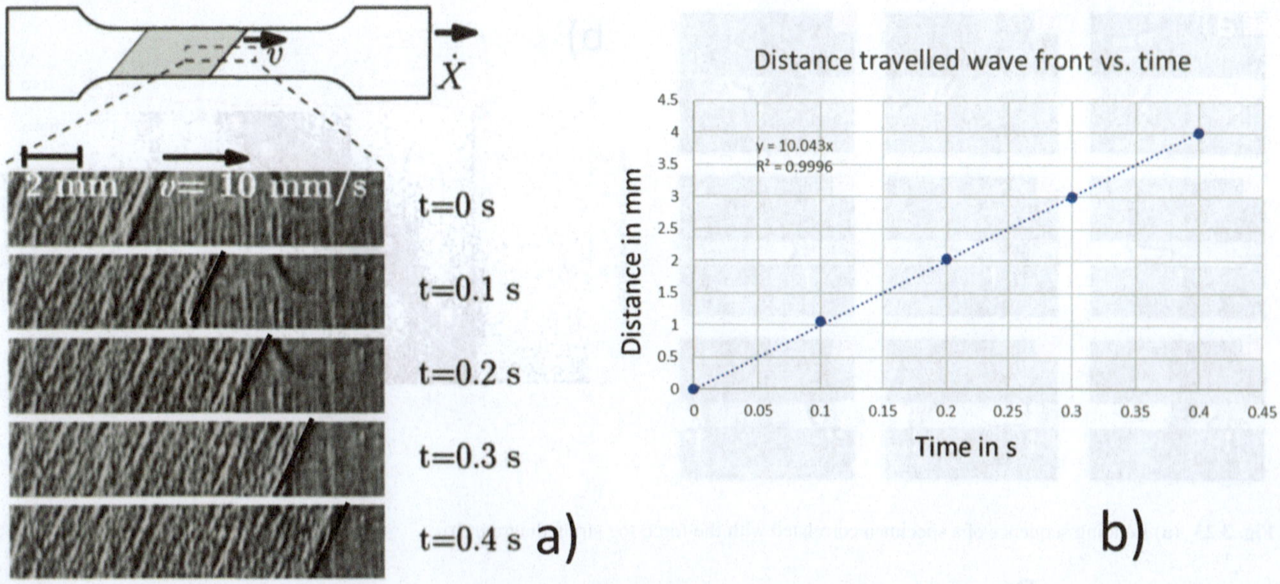

Fig. 3.22 (**a**) Visualization of the propagation of the plastic wave instability using a shadowgraph technique [24]; (**b**) Wave front travelled distance vs. time

The specimen represented in Fig. 3.22 is made of aluminum 5052 similar to the material of the specimen shown in Fig. 3.11 and hence has a f.c.c. crystalline structure. The angle of inclination of the wave front is 25°. The specimen gage length is $L_o = 102$ mm, width $w_o = 12.7$ mm, thickness $t_o = 1$ mm, ratio width to thickness $r_{wt} = w_o/t_o = 12.7$, and pulling speed of the machine, $v_{pm} = 127$ μm/s. For comparison purposes, we recall those for the specimen of Fig. 3.11: $r_{wt} = w_o/t_o = 12.5$, pulling speed of the machine, $v_{pm} = 5.83$ μm/s, that is both r_{wt} are very close, but the pulling speed of this specimen is almost 22 times greater that the speed applied to the specimen of Fig. 3.11. From Fig. 3.22a, it is possible to correlate the space covered by the plastic wave front with the elapsed times giving a speed of the plastic wave equal to 10 mm/s that agrees with the value indicated in Fig. 3.22. The method applied to measure speed in [24] is based in the visual change of roughness of the specimen surface, and the size of the wave front pulse is not measurable in Fig. 3.22. This speed is measured in a frame attached to the specimen, the specimen is moving with a speed of 0.127 mm/s, that is, the ratio of the velocities is $r_{vl} = 10/0.127 = 78.7$. For comparison purposes, the speed of sound in aluminum is 6.3×10^6 mm/s, many orders of magnitude larger than the propagation of the plastic wave front. In [24], a model is presented that deals with the propagation of a plastic front at constant stress, that is, at zero stress rate, and the conclusion of the model based on experimental measurements is that

$$\mathrm{v_{pm}} = \mathrm{v_{wB}} \times \Delta\varepsilon_p \tag{3.24}$$

In Eq. (3.24), $\Delta\varepsilon_p$ is the jump on strain across the plastic wave front, and, according to experimental measurements, $\Delta\varepsilon_p$ is roughly constant and can change in order of magnitude if the stress level changes.

To clarify the process of initiation and propagation of the plastic wave front, we can look at the results corresponding to a specimen of the same material than specimens Nos. 1 and 2, [11], but with following dimensions, $L_g = 65$ mm, $w_o = 10$ mm, $t_o = 5$ mm, $v_{pm} = 0.333$ μ/s.

Figure 3.23 shows the propagation of the plastic instability from the onset of plasticity frame 1 to frame 9 close to the fracture. The blue points represent the acoustic events related to the dislocation motions captured by the two acoustic emission sensors (A.E.). The fringe patterns are iso-derivatives fringes as previously shown in the paper (Eqs. (3.18) and (3.19)).

The specimen is loaded with a special testing machine that pulls both ends symmetrically and keeps the center of the specimen in a fixed position. While frame No. 1 exhibits what in the literature is called a wide band, the rest of the frames show patterns of iso-derivatives that are very complex and are indicative of the heterogeneous structure of the specimen influencing the local patterns of deformation, showing reversal of the inclination of the wave front (frames 5 and 6). Frames 7, 8, and 9 show iso-derivatives similar to iso-derivative patterns of Figs. 3.14 and 3.17. These observed patterns are transient and random.

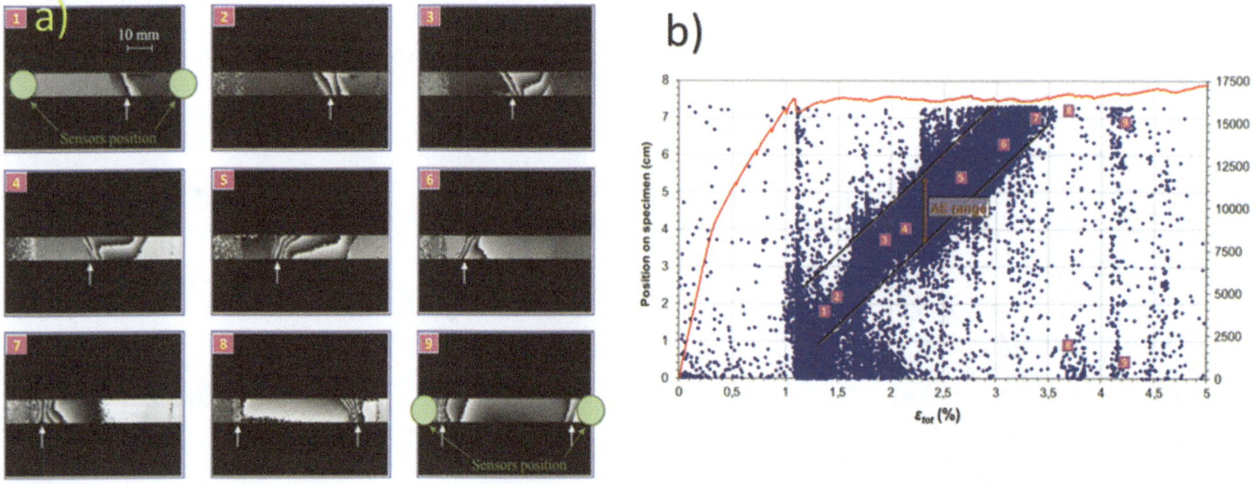

Fig. 3.23 (a) Loading sequence of a specimen correlated with the force-log strain diagram (b)

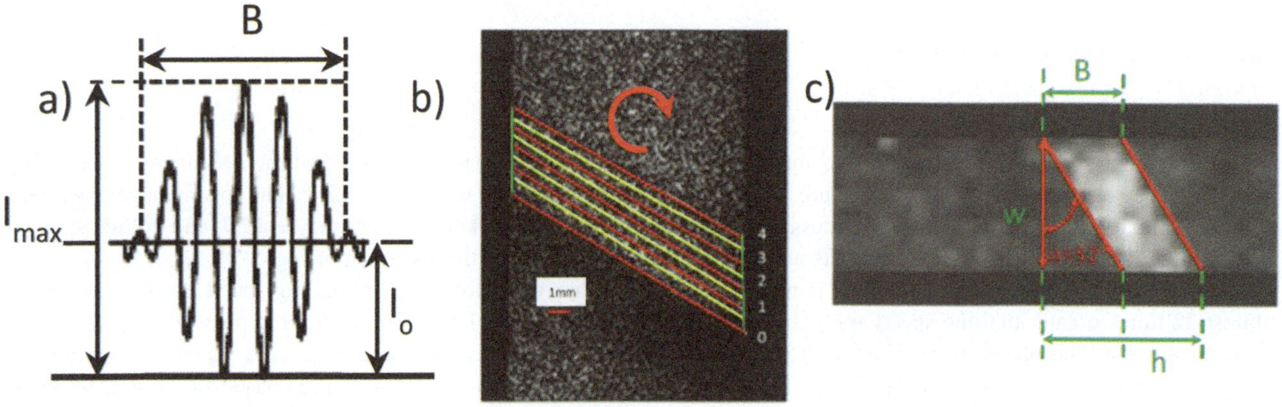

Fig. 3.24 (a) Optical signal; (b) image of the iso-derivatives captured by a CCD camera; (c) low-resolution picture corresponding to the iso-derivatives

3.12 Interpretation of the Optical Signal and Consequences of this Analysis

Having reviewed literature experimental evidence on the propagation of the plastic instability, we will proceed to extract the basic information from the experimental data, first proceeding to analyze the optical signals generated by a wave front and second determining implications of these signals in the analysis of the propagation of the plastic instability taking into consideration the method used to measure velocities.

First, let us consider the shape signal that is propagating along the specimen.

Figure 3.24a represents a wave shape of the signal introduced by the moving end of a testing machine at the onset of plasticity. The signal is a frequency and amplitude modulated signal, the background intensity of the signal is I_0 and the maximum amplitude is $I_{\max}$. The pattern has amplitude and frequency modulated fringes that merge to the background intensity that is not constant but may have the components of very low frequency and amplitude static fringes. The actual shape of the signal depends on the interaction of the specimen and the machine compliance [24]. Figure 3.24b shows the iso-derivatives of a propagating wave front recorded by a CCD camera [25]. The material of the specimen is an aluminum alloy AA5083-0; the dimension of the specimen are $L_g = 25$ mm, $w_o = 10$ mm, $t_o = 3$ mm, 0.5×10^{-2} mm/s. From Fig. 3.24b, the value for the right side of the specimen is $B = 4$ mm and for $h = 10$ mm and applying Fig. 3.24c,

$$B = h - w_o \times tg\alpha \tag{3.25}$$

With $h = 10$ mm, $\alpha = 31°$ one obtains $B = 4$ mm. The sensitivity of the interferometer is $S_y = 0.562$ μm. The red lines represent the minima of the iso-derivatives fringes and the yellow lines the maxima. If we compute the average value of the derivatives for the right side of the specimen, Fig. 3.24b, it results:

$$\varepsilon_{vavg} = \frac{4 \times 0.562}{4000} = 5.62 \times 10^{-4} \tag{3.26}$$

Doing the same computation for the right edge, one obtains 5.37×10^{-4}. This reduction corresponding to a bending moment indicated by the red symbol of Fig. 3.24b.

In the snapshots recorded by a camera, Figs. 3.14 and 3.17, we get the iso-derivative fringes that provide the dimensions of the corresponding signals. In the low spatial and temporal resolution images (see Fig. 3.11), one can see an envelope of the signal that appears as band with a width B, Fig. 3.24b. This signal is inclined with respect to the axis of symmetry of the specimen due to the presence of a bending effect analyzed in the description of the signals of Fig. 3.17. If one looks closer to Fig. 3.24b, the width in the right side (specimen vertical) is smaller than the width in the left side indicating higher strain in the left side.

The next step is to review the experimental procedures to define the velocity and shape of the propagating wave front.

When the moving end of the tensile specimen at the onset of plasticity is displaced, a tensile stress pulse propagates with a finite speed toward the stationary end of the specimen. The average plastic strain produced by the pulse is

$$\Delta\varepsilon_p = \frac{\text{extension}}{\text{space traveled by pulse}} = \frac{v_{pm} \times t}{v_{wB} \times t} \tag{3.27}$$

The space traveled by the pulse will vary with the increment of the stress according to Eq. (3.27) and consequently the strain rate will depend on the shape of the stress-strain curve. Only if the stress-strain plot is flat, $\Delta\varepsilon_p$ will be constant, conclusion that was arrived in [24] in the analysis of the image of Fig. 3.22. In Fig. 3.22, as previously pointed out, the method utilized to measure the speed of the wave front is connected with the change of texture of the surface of the specimen as captured by the camera frame rate.

In [11], to compute the wave front speed v_{wB}, a model is adopted, and this model implies assuming a statistical distribution of the velocity average that includes the band-width B as defined in Fig. 3.24a–c. The position of the wave front is assumed to be defined by the effect of the wave front on the iso-derivatives of the captured images. However, in the paper no distinction between iso-derivatives as defined in this paper and the isothetic lines is made. As shown in Fig. 3.23a, the passage of the wave front at a given location along the gage length of the specimen has very different effects. This fact implies that the energy carried by the wave front is spent in different molecular arrangement processes. This means that velocity changes according to Eq. (3.24). In Refs. [10, 11], the position of wave front is correlated with noise signals produced by the dislocation motion that are complemented with the iso-derivative patters captured by a CCD camera for the selected exposures times. The model adopted in [11] is graphically illustrated by Fig. 3.24b and by Eq. (3.25).

In Eq. (3.25), there are two unknown values, h and α. The value of h is determined from two different sources of information, an estimate of h obtained from Fig. 3.23a is $h = 19.6$ mm. No description is provided for the procedure to obtain h from the patterns of Fig. 3.23a; α is determined also from Fig. 3.23a, $\alpha = 32°$. With these values applying Eq. (3.25), one obtains $B = 13$ mm. In [11], two additional values of B are given, one by averaging measurements along the central line of the specimen used a procedure not described in the paper that yields a value of 14.7 mm, and a third one using AE to measure h that gives a value of $B = 16.5$ mm. The average value of these three measurements gives $B = 14.73 \pm 1.43$, that is, a 9.7% variation.

The validity of the derivation expressed by Eq. (3.24) in agreement with the conclusion arrived in [24] was assumed by P. Hähner [26] that introduced the concept of a soliton wave, a wave that moves at constant speed without changing shape. In [11], a correction to Eq. (3.24) is introduced

$$\Delta\varepsilon_p = \frac{v_{pm}}{v_{Bw}}\left(1 + \frac{B}{L_g}\right) \tag{3.28}$$

The reason of this correction, that is not meaningful for short specimens where $B/L_g << 1$, deals with the fact that at a given section of the specimen the wave front has to pass head to tail to leave a constant $\Delta\varepsilon_p$.

Looking back to Fig. 3.11, one can see that in the low spatial and time resolution image, we see the plastic wave front advancing along the specimen with the value of B that is practically constant indicating a soliton wave propagation.

Figure 3.24c corresponds to the specimen of Fig. 3.11 and the measured $B = 13$ mm. For this specimen, $w_o = 25$ mm, the ratio $B/w_o = 0.52 \pm 0.2$ and since for the specimen of Fig. 3.14, it holds $L_n \approx 2w_o$, the relation is valid also for the current specimen leading to $B \approx 0.25\ L_n$. From Fig. 3.24b, one obtains $B/w_o = 0.50 \pm 0.3$. This result means that this ratio is geometrically determined and independent of the speed of propagation.

3.13 Discussion and Conclusions

The subject matter of this paper has a sizeable amount of papers that deal with it [26] and a variety of interpretations of different aspects of experimental and theoretical approaches. The methodology in this paper has been to utilize an Experimental Mechanics tool and the iso-derivative fringes (moiré method) via speckle interferometry to obtain information about the processes that arise from the onset of plastic instability in tensile specimens with rectangular cross sections. This process has an initiation of the instability, followed by a transition to fracture.

The first step in the analysis of the problem is the characterization of the displacement field in a tensile specimen in the elastic range and then determining the effect of the onset of plasticity in the displacement field. After briefly reviewing the scale effect on recordings of processes, it introduced a family of fringes and the iso-derivatives fringes, and the properties of these fringes are outlined. The transition from the elastic behavior to the onset of plasticity is illustrated with an example showing that this transition can be pinpointed without ambiguity in a sequence of frames of a movie recording of the process.

The next step is the analysis of what happens after the onset of plastic instability, the transition between plasticity and final fracture. This is a more difficult problem that in the current state of the literature and in spite of the very many papers published is not fully explained.

The concept of instability of the plastic flow is reviewed, and an illustrative simplified example of a particular case is shown connecting dislocation motion with a feature that is observed in optical recordings, a wide band, either directly with isothetic fringes or through computation for example using digital image correlation (DIC).

In the analysis of the plastic instability, an important aspect is the geometrical changes that take place in the specimen that are contained in the concept of necking with two variants diffuse necking and necking. Both concepts are experimentally illustrated with the utilization of iso-derivatives applied to examples from the literature. The diffuse necking that corresponds to the propagation or formation of the instability, necking leading to the fracture.

The next stage is to consider the dynamic effect associated with the transition from onset of plasticity to final fracture. The plastic instability is associated with a wave front that moves with a speed depending on the speed of the application of displacements by the testing machine. Examples from the literature are utilized to evaluate the relationship between the rate of application of the displacement of the machine and the velocity of the wave front.

Since the propagation of a plastic wave front is associated with the presence in the image of wide bands, the first step is to establish the optical format of the signal, the effect of the time and space resolutions in the recorded of the images to analyze different cases. In the high spatial and time resolution examples, characteristics of necking process, whether diffuse or non-propagating, are provided by the iso-derivatives, and it shows the influence of the molecular structure of the metal, which according to the available slip lines and the presence of other factors determines the position of the axial resultant in the fracture mode that in the simplified 2D plane stress case results in a mode-1 fracture or when the position resultant force produces bending stresses that result in a mixed mode fracture mode-1 and mode-2. All the analyzed examples correspond to f.c.c metals, and this explains the approximately equal characteristic angle of the fracture processes.

In the case of low spatial and time resolution, the wide bands are characterized by two parameters, the parameter B characterizing the width of the signal, and the angle α that, as indicated before, is connected to the presence of bending in the plane of the specimen. In the paper, it is concluded that B mainly depends of geometric factors and is independent of the speed of propagation.

One of the most discussed factors in the literature is the propagation part of the wave front and the source of many different arguments and proposed "classifications" and nomenclatures. Let us first address the propagation problem. There are two main types of plastic instabilities: (1) non-propagating and (2) propagating. An example of non-propagating instability is illustrated by the specimen of Figs. 3.7 and 3.11, the geometry of the specimen determines where the onset of fracture will take place. After the local necking starts, the process of fracture will be similar to that of the plastic analysis of structures [27]. The other case is the presence of defects of different types in the metal that determine where the plastic instability manifests and leads to fracture. The most complex case to discuss is the propagating case, where the variety of possible outcomes is practically innumerable. Also, the use of the nomenclature is confusing, the name of Luders' bands or shear bands applies to all cases of propagation in the sense that the passage of the plastic wave implies an increment of the roughness of the surface of a specimens as shown in Fig. 3.22. We have analyzed this process and documented the roughness change in the plastic range of tensile specimens as the specimen is stretched in continuous stretching or periodic stretching in the case of fatigue. A very interesting case of wide band formation is the bands associated with the Portevin-Le Chatelier effect that manifests in the form of serrations in the stress-strain plot that appear at a critical strain and is both temperature and strain rate dependent. These serrations have variable amplitude and frequencies.

To clarify ideas, it is interesting to analyze the case of the soliton wave illustrated in Fig. 3.11. As shown in this paper, and is supported in the literature, a soliton wave can only take place if the metal tested has a very homogeneous structure that satisfies the condition $\frac{d\sigma}{d\varepsilon} = 0$. In Fig. 3.11, an experimental observation of this propagation is provided within the approximation of low spatial and time resolution. Because of the required conditions in the structure of metal, the experimental observation of a soliton wave is not a very common event. However, it was the starting point for the study of the propagation of plastic waves by Hähner [28], later followed by Yoshida [6]. The analysis of Fig. 3.23 provides the most probable type of propagation that can be observed influenced by local structure changes and different plastic instabilities possible in the wave propagation. In [11], a statistical approach based in the correlation of optical signals and AE signals is utilized to get an average value of the plastic wave propagation velocity. The passage of the wave creates different iso-derivatives at particular positions. Correlating these patterns with AE signals provides average speed of propagation that is utilized to relate the velocity of propagation of the signal optically detected with the speed of motion applied to the machine. A correction is introduced to take into consideration the fact that the plastic wave front propagates in a finite gage length.

There are many additional developments that are required for fully comprehend all the different and relevant phenomena involved in the topics deal within this paper. One of the possible practical applications is to develop an accurate and quick way to evaluate the quality of materials fabricated using additive manufacturing.

References

1. Panin, V.E.: Physical Mesomechanics of Heterogeneous Media and Computer Aided Design of Materials. Cambridge Interscience Publishing, Cambridge (1998)
2. Yoshida, S., Muchiar, I., Muhamad, R., Widiastuti, R., Kusnovo, A.: Optical interferometric technique for deformation analysis. Opt. Expr. **2**(13), 516–530 (1998)
3. Yoshida, S., Siahaan, B., Pardede, M.H., Sijabat, N., Simangunsong, H., Simbolom, T., Kusnovo, A.: Observation of plastic deformations wave in a tensile-load aluminum alloy. Phys. Lett. A. **251**(1), 54 (1999)
4. Yoshida, S., Rourks, R.L., Mita, T., Ichinose, K.L.: Physical mesomechanical criteria of plastic deformation and fracture. Phys. Mesomech. **12**(5–6), 249–253 (2009)
5. Yoshida, S.: Physical mesomechanics as a field theory. Phys. Mesomech. **8**(5–6), 15–20 (2005)
6. Yoshida, S.: Comprehensive description of deformation of solids in wave mechanics. Math. Mech. Compl. Sys. **3**(3), 243–272 (2015)
7. Durelli, A.J., Parks, V.J.: Moire Analysis of Strain. Prentice-Hall, Englewood Cliffs (1970)
8. Petit, J.: Elastic-plastic behavior and strain localization – multi-scale study by ESPI and acoustic emission. Ph.D. Thesis, University of Technology of Troyes, France (2010)
9. Petit, J., Montay, G., Francois, M.: Localization phenomenon investigation on stainless steel samples by Speckle interferometry. Strain. **47**, 363–371 (2011)
10. Petit, J., Montay, G., Francois, M.: Strain rate measurements by speckle interferometry for necking investigation in stainless steel. Int. J. Solids Struct. **51**(2), 540–550 (2014)
11. Petit, J., Montay, G., Francois, M.: Strain localization in mild (low carbon) steel observed by acoustic emission – ESPI coupling during tensile test. Exp. Mech. **58**, 743–758 (2018)
12. Hill, R.: Elastic properties of reinforced solids: some theoretical principles. J. Mech. Phys. Solids. **11**(5), 357–372 (1963)
13. Mandel, J.: Plasticité classique et viscoplasticité: CISM-1971. Springer, Udine (1972)
14. Bazant, Z.P.: A correlation study of formulations of incremental deformation and stability of continuous bodies. J. Appl. Mech. **38**, 919–928 (1971)
15. Bazant, Z.P., Cedolin, L.: Stability of Structures. World Scientific Publishing Co., Singapore (2017)
16. Bazant, Z.P., Gattu, M., Vorel, J.: Work conjugacy error in commercial finite-element codes: its magnitude and how to compensate for it. Proc. R. Soc. A. **468**, 3047–3058 (2012)
17. Bazant, Z.P., Vorel, J.: Energy-conservation error due to use of Green–Naghdi objective stress rate in commercial finite-element codes and its compensation. J. Appl. Mech. **81**, Paper No. 021008 (2014)
18. Bazant, Z.P., Adley, M.D., Carol, I., Mirásek, M., Akers, S.A., Rohani, B., Cargile, J.D., Caner, F.C.: Large-strain generalization of micro-plane model for concrete and application. J. Eng. Mech. **126**, 971–980 (2000)
19. Bazant, Z.P., Le, J.-L.: Probabilistic Mechanics of Quasi-Brittle Structures. Cambridge University Press, Cambridge (2017)
20. Sciammarella, C.A., Lamberti, L., Sciammarella, F.M.: Verification of continuum mechanics predictions with experimental mechanics. Materials. **13**, Paper No. 13010077, 1–25 (2020)
21. Sciammarella, C.A., Sciammarella, F.M.: Experimental Mechanics of Solids. Wiley, Chichester (2012)
22. Horgan, C.O., Payne, L.E.: A Saint-Venant principle for a theory of non-linear plane elasticity. Q. Appl. Math. **50**, 641–675 (1992)
23. Cordero, R., Martinez, A., Rayas, J.A., Labbe, F.: Necking progression in tensile specimens monitored in real-time by using fringe projection. Opt. Lasers Eng. **48**, 1285–1290 (2010)

24. Mertens, F., Frankling, S., Marder, M.: Phys. Rev. Lett. **78**(23) (1997)
25. Nakamura, T., Sasaki, T., Yoshida, S.: Analysis of Portevin-Le Chatelier effect of Al-Mg alloy by electronic speckle interferometry. In: Advancement of Optical Methods in Experimental Mechanics, vol. 3. Conference Proceedings of the Society for Experimental Mechanics Series (2014)
26. Yilmaz, A.: Portevin-Le Chatelier effect: a review of experimental findings. Sci. Technol. Adv. Mat. **12**, Paper No. 063001, 1–16 (2011)
27. Hodge, P.G.: Plastic Analysis of Structures. McGraw-Hill Book Company, New York (1959)
28. Hähner, P.: Theory of solitary plastic waves – part I: Lüders bands in polycrystals. Appl. Phys. A Mater. Sci. Process. **58**, 41–48 (1994)

Chapter 4
A Digital Laser Speckle Technique for Generating Slope, Curvature, and Deflection Contours of Bent Plates

Austin Giordano and Fu-Pen Chiang

Abstract A digital method is proposed whereby spatial speckles created when an optically rough surface is illuminated by a diffuse coherent laser beam are used to generate slope contour fringes. This is done by photographing the speckles contained in a parallel plane in front of the plate before and after deformation. The resulting photographs are then fast Fourier transformed twice and numerically compared with the digital image correlation (DIC) software computer-aided speckle interferometry (CASI) to produce slope contours of the plate. The slope contour data is then stored in an array. The curvature contours are generated by numeric differentiation of the slope data; similarly the deflection contour is generated by numeric integration of the slope data. The curvature data can then be further used in conjunction with plate theory to obtain the bending and twisting moments, the plate's experiences, as well as the fiber stress on the surface of the plate.

Keywords Laser speckle · Digital speckle photography · Digital image correlation · Stress analysis · Plate bending

4.1 Introduction

When a coherent laser beam impinges upon an optically rough surface of a specimen, the reflected wavelets from every point of the surface mutually interfere into a volumetric speckle pattern. These speckles move when the surface of the specimen move, and they tilt when the surface of the specimen tilts in much the same way as a mirror tilts an impinging light beam. These speckles can be used to measure the deformation of the specimen. If a camera is focused at the surface of the specimen, the speckles act as if they are in-plane displacement gages; but when the camera focuses at a plane away from the surface of the specimen, the movement of the speckles can be used to measure the slope of the surface. A detailed description of using laser speckle for deformation measurement is given by Chiang [1]. Laser speckle has been used for metrology and experimental mechanics for some time, but most of the work has been done before digitization. In this paper, we propose a digital technique for slope measurement. Defocused speckles have been used by Juang and Chiang [2] to generate slope contours and by Keene and Chiang [3] to measure vibrational modes. Many experimental methods for solving plate bending problems are methods of deflection measurement [4, 5]; however, in the case of thin plate bending, the more advantageous approach is to generate the slope contour thus eliminating the need to differentiate twice to obtain the curvature of the plate. This would reduce the errors associated with numerical differentiation. Techniques for obtaining the curvature of a plate were presented by Kao and Chiang [6], these approaches required the use of a fine grating to produce the curvatures. Additionally Kao and Chiang's method was incapable of producing the twisting curvature without capturing the curvature of the plate in a rotated coordinate system; no such requirement exists with the digital technique proposed.

In this paper, a full-field technique of slope measurement is described in which the specimen surface can be of almost any texture. Using the slope measurement curvature can simply be obtained via numeric differentiation, and deflection can be obtained by numeric integration. This technique is based on the usage of speckles that are created by a beam of coherent light from a laser that is passed through a microscope objective and a pinhole aperture. With this method, no surface preparation is necessary if sufficient light is reflected back from the surface. Otherwise, a thin coating of white paint sprayed onto the surface is all that is needed.

A. Giordano (✉) · F.-P. Chiang
Department of Mechanical Engineering, State University of New York at Stony Brook, Stony Brook, NY, USA
e-mail: austin.giordano@stonybrook.edu; fu-pen.chiang@stonybrook.edu

M.-T. Lin et al. (eds.), *Advancement of Optical Methods & Digital Image Correlation in Experimental Mechanics*,
Conference Proceedings of the Society for Experimental Mechanics Series, https://doi.org/10.1007/978-3-030-59773-3_4

4.2 Experimental Procedures

The system used for the data acquisition is as shown in Fig. 4.1a. The specimen is illuminated by an expanded laser beam, and the image of the object is captured by a digital camera. The light intensity of each speckle pattern is digitized into an array of 8688 × 5792 pixels. The image processing software ImageJ [7] is then used to convert the JPG image files into 8-bit TIFF files. The 8-bit TIFF files are then loaded into an image correlation software based on Fourier transform [8] called CASI and processed. Basic processes of the technique involve data acquisition and image processing. In the data acquisition stage, two speckle patterns of the specimen, one before and one after the deformation, are captured by the camera and stored on an SD card, which is then registered into the computer. The image processing stage consists of four steps with ImageJ and CASI. First, the JPG image files are converted into 8-bit TIFF files which are then input into CASI. The three steps of the process that CASI is necessary for was described by Chen and Chiang [8]. First, an equivalent double-exposure speckle pattern is obtained by superimposing the two digital images, and it is segmented into a series of small subimages. Second, a fast Fourier transform (FFT), is applied to each subimage and "compared" with a correlation calculation. Last, a second FFT is applied to the resulting value giving rise to an impulse function whose position vector is nothing but the displacement vector experienced collectively by all the speckles contained within the subimage. Repeating the last two steps to all other subimages, we can resolve the full-field 2-D displacements. Upon completion of this process, the slope fields are generated. These slope fields are then numerically integrated to produce the deflection of the plate and numerically derived to produce the curvatures of the plate. An image of the algorithm that CASI performs is displayed in Fig. 4.2.

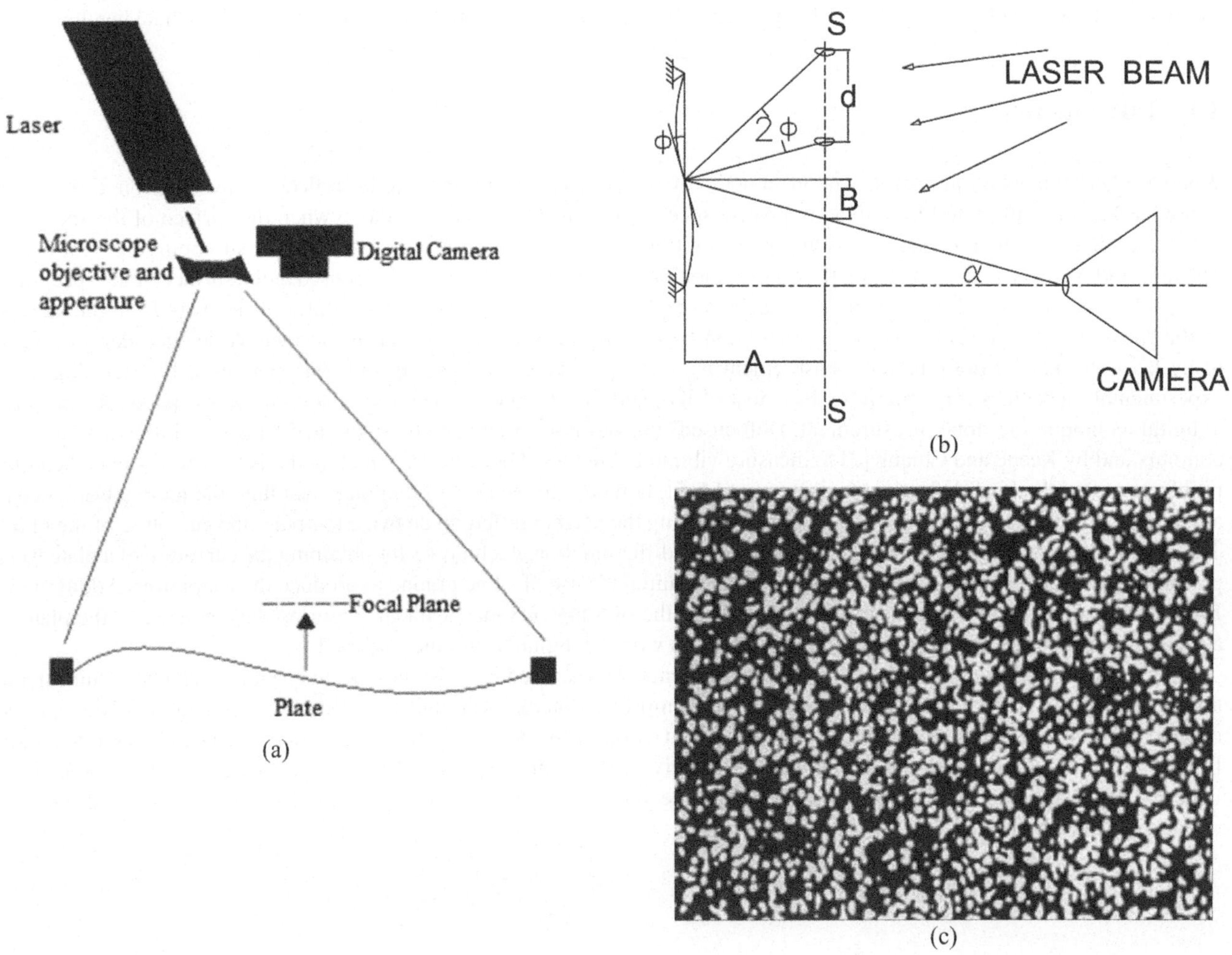

Fig. 4.1 (**a**) Schematic of the data-acquisition system. (**b**) Arrangement for recording spatial speckles for plate bending. (**c**) Sample of black and white speckles

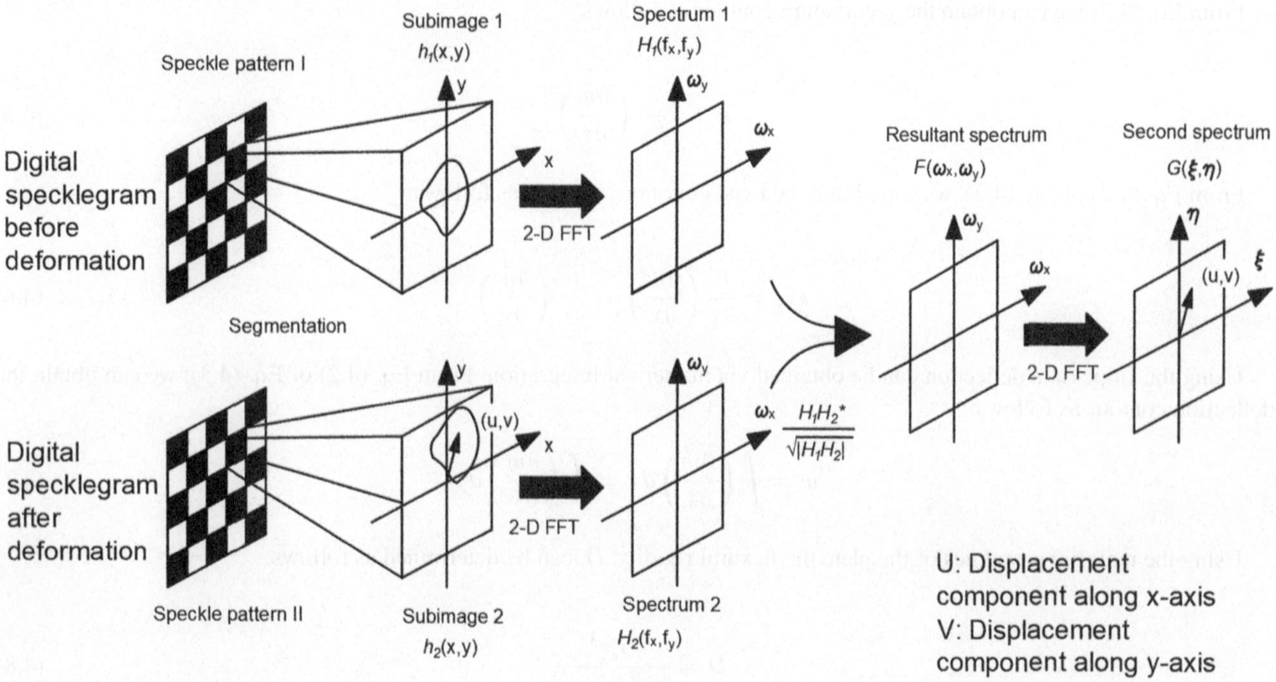

Fig. 4.2 Algorithm by which CASI operates

4.3 Principle of the Method

Fringes are formed when coherent waves interfere. When the number of point sources increases, the interference pattern takes the form of randomly distributed light and dark spots of varying size. These light and dark spots are speckles. The speckles in space will move when the specimen surface is moved. Under the classical assumptions of the thin plate theory, the stress field can be determined once the surface curvature is obtained. The present technique produces slope contours from which curvature can be calculated. By selecting an arbitrary plane in space parallel to the surface of the plate, the speckles contained in the plane will move according to the local surface tilt. The displacement of the speckles d can be related to the surface slope ϕ. From Fig. 4.1b it can be shown that [2]

$$d = \frac{\mathrm{A}\left[\tan\left(2\phi+\alpha\right)-\tan\left(\alpha\right)\right]}{\frac{\tan 2\phi\left(1+\tan^2\alpha\right)}{1-\tan(\alpha)\tan(2\phi)}}. \tag{4.1}$$

From Eq. (4.1), we can obtain the slope contours as follows depending on which direction we select to display the result:

$$\frac{\partial w}{\partial x} = \phi_x = \frac{\mathrm{U}}{2\mathrm{A}}. \tag{4.2}$$

When the input for d is the U-displacement.

$$\frac{\partial w}{\partial y} = \phi_y = \frac{\mathrm{V}}{2\mathrm{A}}. \tag{4.3}$$

When the input for d is the V-displacement.

Using the slope data curvature can be obtained via numerical differentiation. From Eq. (4.2) we can obtain the x-curvature contour as follows:

$$\kappa_x = \frac{\partial}{\partial x}\left(\frac{\partial w}{\partial x}\right). \tag{4.4}$$

From Eq. (4.3) we can obtain the y-curvature contour as follows:

$$\kappa_y = \frac{\partial}{\partial y}\left(\frac{\partial w}{\partial y}\right). \tag{4.5}$$

From Eq. (4.2) or Eq. (4.3), we can obtain the twist curvature contour as follows:

$$\kappa_{xy} = \frac{\partial}{\partial y}\left(\frac{\partial w}{\partial x}\right) = \frac{\partial}{\partial x}\left(\frac{\partial w}{\partial y}\right). \tag{4.6}$$

Using the slope data deflection can be obtained via numerical integration. From Eq. (4.2) or Eq. (4.3), we can obtain the deflection contour as follows:

$$w = \int \left(\frac{\partial w}{\partial x}\right) dx = \int \left(\frac{\partial w}{\partial y}\right) dy \tag{4.7}$$

Using the material properties of the plate the flexural rigidity, D, can be determined as follows:

$$D = \frac{Et^3}{12\left(1 - \nu^2\right)} \tag{4.8}$$

where E is the Young's modulus, t is the plate thickness, and ν is the Poisson's ratio. From the curvatures obtained in Eqs. (4.4, 4.5, and 4.6), and the flexural rigidity, D determined in Eq. (4.8), the following relations exist between the moments per unit length M_x, M_y, and M_{xy}, and the plate curvatures [9]:

$$M_x = -D\left(\kappa_x + \nu\kappa_y\right)$$

$$M_y = -D\left(\kappa_y + \nu\kappa_x\right)$$

$$M_{xy} = -D\left(1 - \nu\right)\kappa_{xy} \tag{4.9}$$

The fiber stresses of the plate due to the bending and twisting moments are given by:

$$\sigma_x = \frac{6M_x}{t^2}$$

$$\sigma_y = \frac{6M_y}{t^2}$$

$$\tau_{xy} = \frac{6M_{xy}}{t^2} \tag{4.10}$$

It is seen from Eqs. (4.9) and (4.10) that the stresses are related to the values of curvatures and twist, the second derivatives of plate deflection. Once the curvatures and the twist are experimentally determined, the stresses can be determined.

4.4 Results

In this experiment, we tested the deformation of a circular disk under a central transverse load using the proposed laser speckle method. The plate was made of polycarbonate (diameter = 10.16 cm, thickness = 6.35 mm) and loaded by a central deflection of 12.7 μm. A 15 mW expanded He-Ne laser beam ($\lambda = 0.6328$ μm) was used to illuminate the object from 2.2 m away. The digital camera was situated 2.1 m away from the plate and focused on a parallel plane 17.5 cm in front of the specimen surface. Figure 4.3 shows a set of slope fringes obtained by processing a before and after loading image with CASI. There are situations in which deflection of a plate is needed, such as the warpage of an electronic package chip caused by different thermal expansion of different materials. These deflections can be simply calculated using a numeric integration of the slope data; such deflection contours are plotted in Fig. 4.4. In the case of determining the stresses, bending, and twisting moments, the curvature contours are needed, they are produced through a numeric differentiation of the slope data, and a set of curvatures are shown in Fig. 4.5. As the contours are numerically determined and all of the data is stored in an array manipulation to determine the stresses, bending and twisting moments becomes rather easy. The central deflection of the plate shown in Fig. 4.4 has a value of 12.75 μm which compares favorably with the actual deflection of 12.7 μm. The curvatures numerically derived have a good agreement with both theory and pervious experiments.

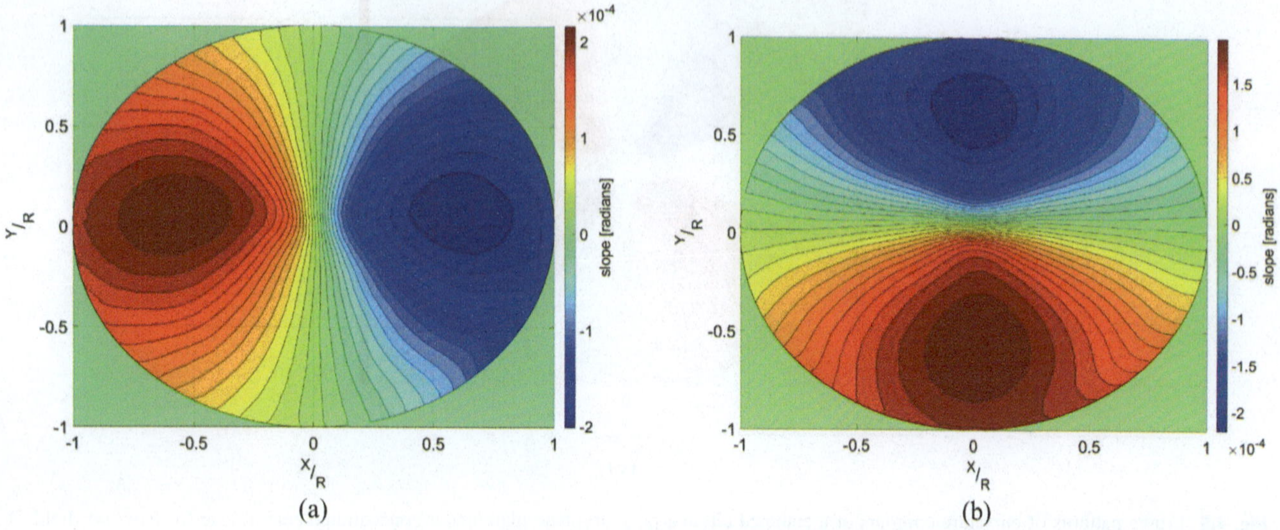

Fig. 4.3 Fringe patterns of partial slope contours of a clamped circular polycarbonate plate under concentrated central load (**a**) $\partial w/\partial x$–field (**b**) $\partial w/\partial y$–field

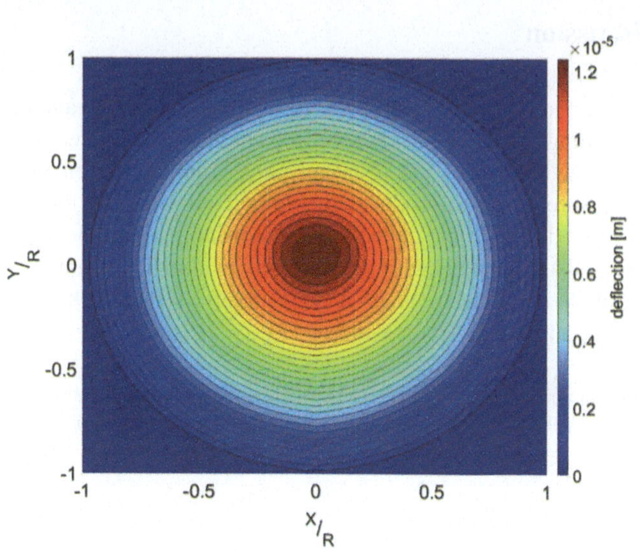

Fig. 4.4 Fringe patterns of deflection contours of a clamped circular polycarbonate plate under concentrated central load

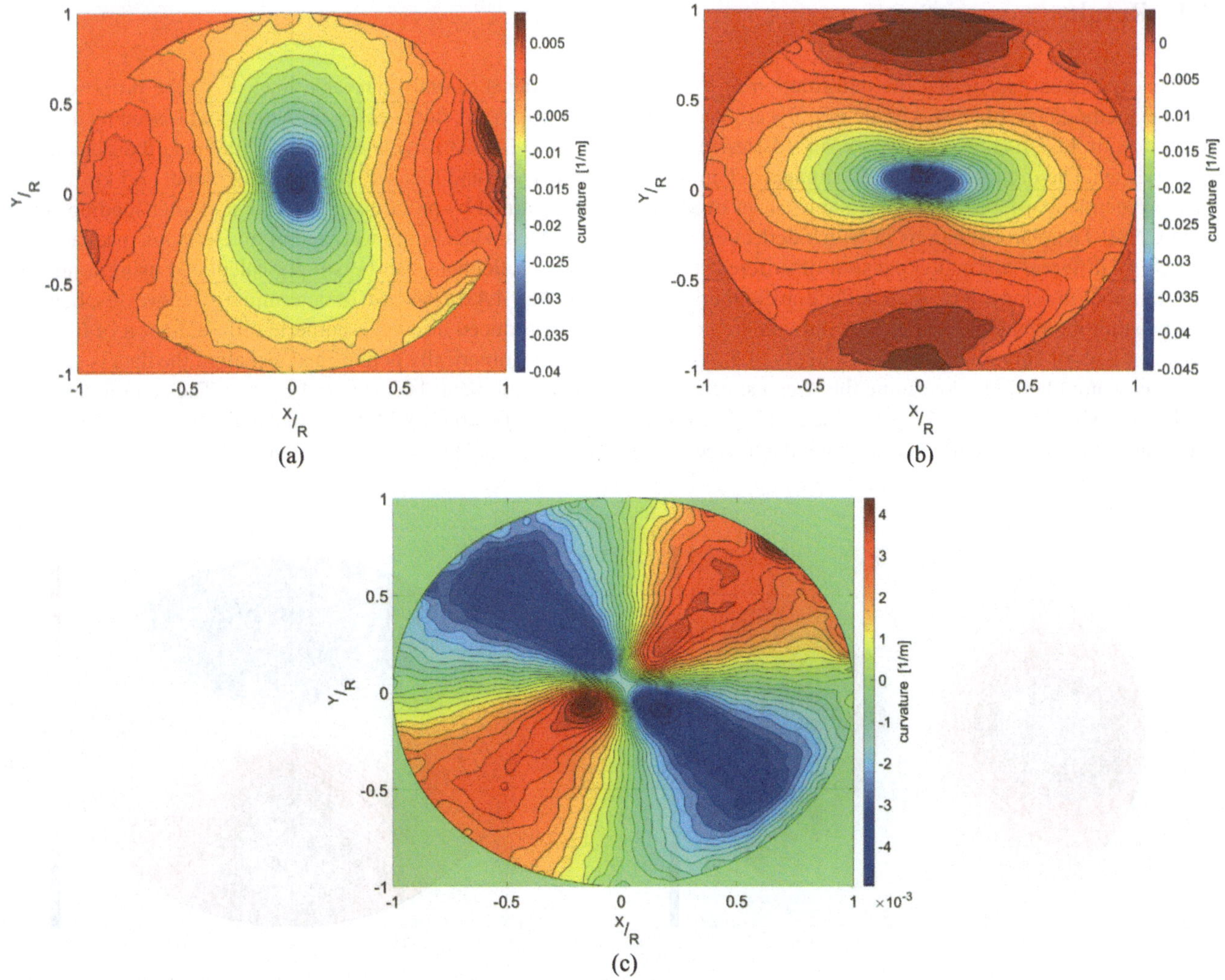

Fig. 4.5 Fringe patterns of curvature contours of a clamped circular polycarbonate plate under concentrated central load (**a**) $\partial^2 w/\partial x^2$–field, (**b**) $\partial^2 w/\partial y^2$–field, (**c**) $\partial^2 w/\partial x \partial y$–field

4.5 Conclusion and Discussion

A digital method has been proposed to use laser speckle generate slope, curvature, and deflection contours of a bent plate. This is achieved by digitally recording the laser speckle pattern form in a plane at certain distance away from the specimen surface before and after loading the plate; and the resulting photographs are then fast Fourier transformed twice with the digital image correlation software CASI to produces the slope contours. The curvature contour is then produced by numerically differentiating the slope data, and the deflection is produced by numerically integrating the slope data. As this technique is a digital technique, the data is stored within an array, which allows for simple manipulation to obtain the stresses, twisting, and bending moment that the plate experiences due to the load. Experience has shown that a thin coat of white paint usually improves a great deal the quality of the fringe pattern.

Acknowledgment The work reported in this paper was partially supported by the Office of Naval Research through award NO.-N00014-17-1-2873. We would like to thank Dr. R.S. Rajapakse, Director of the ONR Solid Mechanics Program, for his support over the years on the development of speckle technology.

References

1. Chiang, F.P.: A family of 2D and 3D experimental stress analysis technique using laser speckle. Solid Mech. Arch. **3**(1), 1–32 (1978)
2. Chiang, F.P., Juang, R.M.: Laser speckle interferometry for plate bending problems. Appl. Opt. **15**, 2199–2204 (1976)
3. Keene, L., Chiang, F.P.: Real-time anti-node visualization of vibrating distributed systems in noisy environments using defocused laser speckle contrast analysis. J. Sound Vib. **320**, 472–481 (2009)
4. Theocaris, P.S.: Moire method in plates. In: Proceedings of the International Association of Shell Structures, pp. 877–889. Symposium, Warsaw (North Holland, Amsterdam) (1963)
5. Havanesian, J., Varner, J.: Method for determining the bending moments in normal loaded thin plates by holographic interferometry. In: Robertson, E.R., Harvey, J.M. (eds.) Engineering Uses of Holography, pp. 173–185. Cambridge University Press, London (1970)
6. Kao, T.Y., Chiang, F.P.: Family of grating techniques of slope and curvature measurements for static and dynamic flexure of plates. Opt. Eng. **21**(4), 214721 (1982)
7. Schneider, C.A., Rasband, W.S., Eliceiri, K.W.: NIH image to ImageJ: 25 years of image analysis. Nat. Methods. **9**, 671–675 (2012)
8. Chen, D.J., Chiang, F.P.: Computer-aided speckle interferometry using spectral amplitude fringes. Appl. Opt. **32**, 225–236 (1993)
9. Timoshenko, S., Woinowsky-Krieger, S.: Theory of Plates and Shells, 2nd edn. McGraw-Hill, New York (1959)

Chapter 5
Holography and Holographic Interferometry via Photopolymer Film

Austin Giordano, Lionel T. Keene, Ryan Norris, and Fu-Pen Chiang

Abstract A new method of performing holography and holographic interferometry is proposed, whereby the film does not have to be developed in the manner that traditional film needs to be developed. This is done by implementing a photopolymer film, a film that develops via a reaction diffusion-driven photo-polymerization process. The utilization of photopolymer film rapidly simplifies the process of performing holography and holographic interferometry as it removes the wet development process that was previously required in holography.

Keywords Holography · Holographic interferometry · Photopolymer · Plate bending · Displacement contours

5.1 Introduction

Holographic film has the capability to store the interference pattern of waves of light. Once developed using a reference beam, the holographic film can reconstruct the image of the object that was used to create the hologram. Many have applied holography and holographic interferometry to study mechanics [1–3]. A wide variety of materials have been implemented in holography [4, 5] recently; a new holographic film material has been developed by Bayer named Bayfol HX photopolymer [6]. Bayfol HX photopolymer has allowed for a simpler process of producing holograms. Bayfol HX photopolymer produces a real-time hologram due to the development process; high index monomers dissolved in the PUR matrix polymerize when exposed to a holographic interference pattern. After the monomers have linked together to produce the fringes of the hologram, the film is exposed to white light to fix and bleach the film and complete the development procedure. This process removes the difficulty associated with the wet chemistry used in the development of silver halide film traditionally used in holography. The live development process of photopolymer film allows for the film to be exposed to light until it is fully developed; due to the nature of the monomers, there is no way in which the film can be overdeveloped. Photopolymer film additionally allows for the film to be left in place during development, and then the object can be deformed or moved to perform holographic interferometry.

In this paper, usage of the new photopolymer material to obtain holograms and performing holographic interferometry is presented. A simple plate bending is used to depict the capability of the new type of film.

5.2 Experimental Procedures

The system used for the data acquisition is as shown in Fig. 5.1. A laser beam is split into two separate beams, the reference and object beam, using a beam splitter. Then the object beam is expanded and illuminates the specimen of interest; simultaneously the reference beam is expanded and illuminates the holographic film. The laser power is tuned using a

A. Giordano (✉) · R. Norris · F.-P. Chiang
Department of Mechanical Engineering, State University of New York at Stony Brook, Stony Brook, NY, USA
e-mail: austin.giordano@stonybrook.edu; fu-pen.chiang@stonybrook.edu

L. T. Keene
Nondestructive Characterization Institute, Lawrence Livermore National Laboratory, Livermore, CA, USA

M.-T. Lin et al. (eds.), *Advancement of Optical Methods & Digital Image Correlation in Experimental Mechanics*,
Conference Proceedings of the Society for Experimental Mechanics Series, https://doi.org/10.1007/978-3-030-59773-3_5

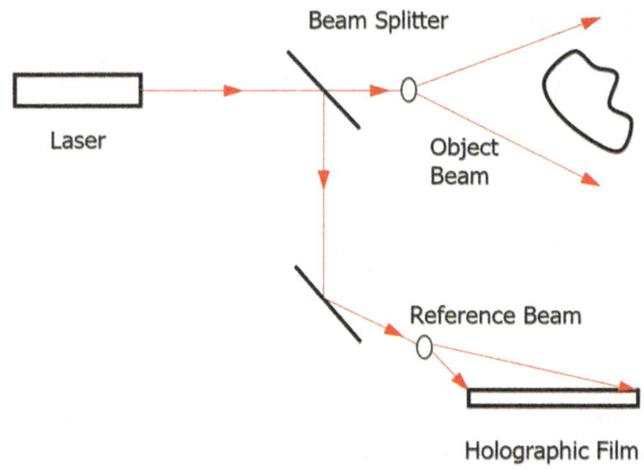

Fig. 5.1 Optical arrangement for producing an off-axis hologram

variable beam splitter to distribute the power sufficiently for the object and film. Setting the power distribution to be about even on each surface of the holographic film, the film can be uncovered and the film to develop. The monomers link together and developing the film and producing the fringes on the film. After the time for the film to develop has passed, the object beam can be blocked to check the reconstruction of the object. The film is then "bleached" with white light and allows any remaining monomers to link with the polymer chains previously created. The object is then deformed, and the interference fringes are then observed. A digital camera is used to image the object through the film to capture the fringes produced by the interference of the new object with the image of the object stored on the holographic film.

5.3 Principle of the Method

Fringes are formed when coherent waves interfere. Holographic interferometry fringes are associated with the phase change of the object to the image of the object stored on the holographic film. This phase change has been studied for out of plane displacement and the relationship which relates phase change, ϕ, wavelength, λ, the incident angle of the object beam, θ, and the out of plane deflection, w, is as follows [7]:

$$\Delta\phi = \frac{2\pi}{\lambda}\left[w\left(1+\cos\theta\right)\right] \tag{5.1}$$

From Eq. (5.1) we can obtain the deflection as follows:

$$w = \frac{\lambda\Delta\phi}{2\pi\left(1+\cos\theta\right)}. \tag{5.2}$$

5.4 Results

In this experiment, we tested the deformation of a circular disk under a central transverse load using the photopolymer film. The plate was made of polycarbonate (diameter = 10.16 cm, thickness = 6.35 mm) and loaded by a central deflection of 12.7 μm. A 45 mW expanded He-Ne laser beam (λ = 0.6328 μm) was used to illuminate the object and film. Figure 5.2 shows a set of fringes obtained by loading the specimen to various displacements. The displacement fringes displayed in Fig. 5.2 are in good agreement with previous experiments and the plate theory developed by Timoshenko [8].

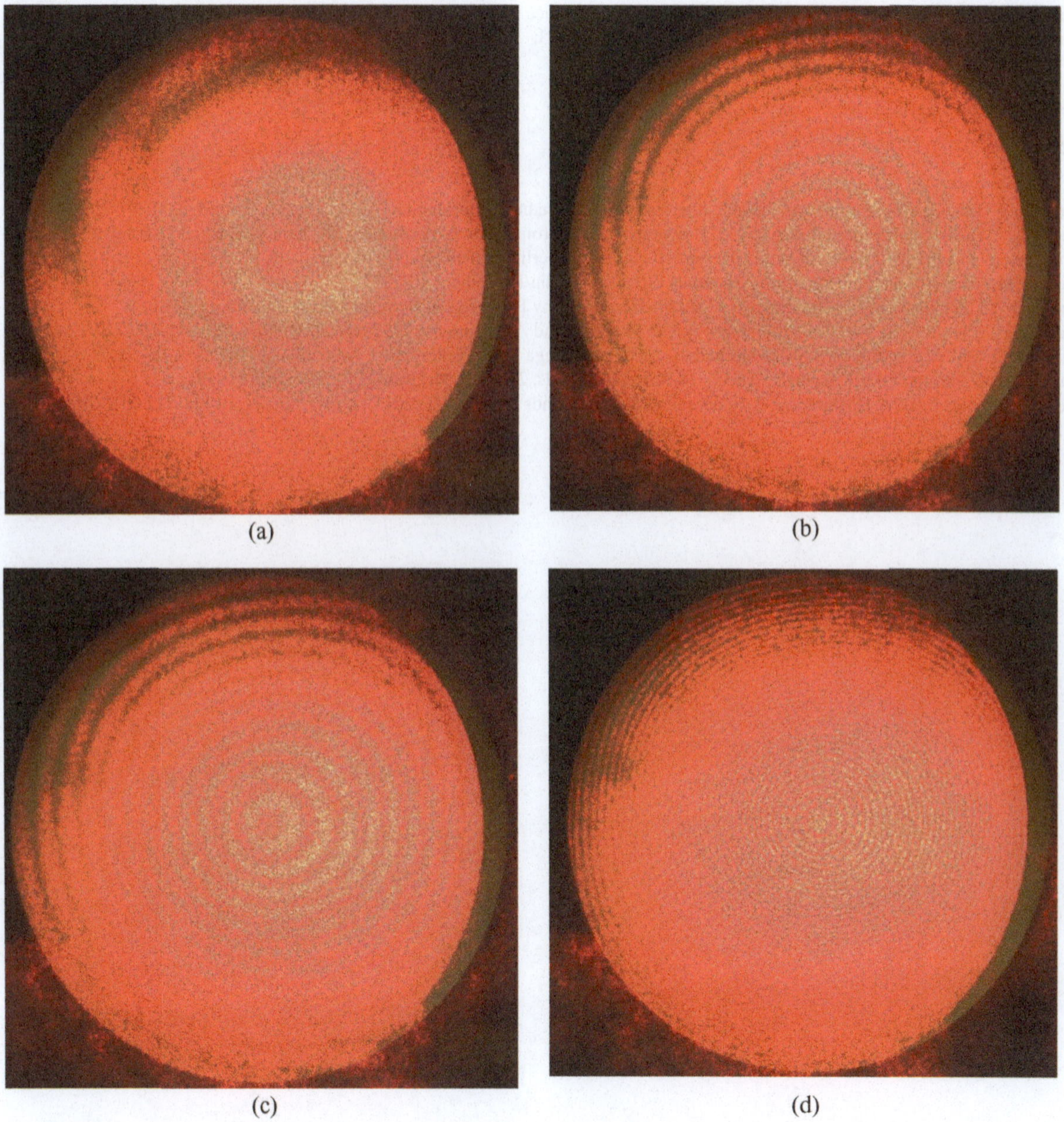

Fig. 5.2 Fringe patterns of a clamped circular polycarbonate plate under concentrated central load. (a) 0.25 μm central displacement, (b) 0.75 μm central displacement, (c) 1.0 μm central displacement, (d) 2.1 μm central displacement

5.5 Conclusion and Discussion

A photopolymer film has been proposed to be used for holography and holographic interferometry. The film greatly simplifies the process of producing holograms as the film cannot be overdeveloped and does not require the wet chemistry and processing of the film. The film developing in place allows for holographic interferometry to be performed simply, by not requiring the film to be removed to be developed. The time required for the development of hologram takes a comparable time to the exposure time of traditional holographic films. Bayer has shown that high beam ratios are sufficient [9], but experience has shown that the beam ratios should be tuned to have approximately equal power on the surface of the film to allow for the best results.

Acknowledgments We gratefully acknowledge the support of Lawrence Livermore National Laboratory, Dr. Rajapakse, and the financial support of the Office of Naval Research through award NO.-N00014-17-1-2873.

References

1. Jones, R., Wykes, C.: Holographic and Speckle Interferometry, 2nd edn. Cambridge University Press, Cambridge (1989)
2. Ostrovsky, Y.I., Shchepinov, V.P., Yakovlev, V.: Holographic Interferometry in Experimental Mechanics. Springer, Berlin/Heidelberg (1991)
3. Schumann, W., Dubas, M.: Holographic Interferometry. Springer, Berlin/Heidelberg (1979)
4. Smith, H.M. (ed.): Holographic Recording Materials. Springer, Berlin/Heidelberg (1977)
5. Ostrovsky, Y.I., Butusov, M.M., Ostrovskaya, G.V.: Interferometry by Holography. Springer Berlin Heidelberg (1980)
6. Proceedings Volume 7957, Practical Holography XXV: Materials and Applications; 79570H (2011). https://doi.org/10.1117/12.876202
7. Shukla, A., Dally, J.: Experimental Solid Mechanics, 2nd edn. College House Enterprises, Knoxville (2014)
8. Timoshenko, S., Woinowsky-Krieger, S.: Theory of Plates and Shells, 2nd edn. McGraw-Hill, New York (1959)
9. Proceedings Volume 8776, Holography: Advances and Modern Trends III; 877603 (2013). https://doi.org/10.1117/12.2018618

Chapter 6
Evaluating Stresses from Measured Strains in Viscoelastic Body Using Numerical Laplace Transformation

S. Taguchi, K. Takeo, and S. Yoneyama

Abstract This study proposes a method for computing stresses from measured values of in-plane strains in viscoelastic body under plane stress condition. Since Poisson's ratio depends on time and temperature, it is difficult to evaluate stresses from in-plane strains unless Poisson's ratio is treated as a constant. This research focuses on pseudoelasticity in the linear viscoelasticity, and the stresses are computed using a numerical Laplace transformation. Since the relation between a through-thickness strain and in-plane strains is expressed in Laplace domain, Poisson's ratio can be treated as time- and temperature-dependence. The Laplace transformation is performed numerically using a method developed by the authors. The stresses are computed from the through-thickness strains and in-plane strains by numerical integration. Therefore, the stresses are computed ignoring the effect caused by using the constitutive equations in Laplace domain. The effectiveness of the proposed method is demonstrated by computing stresses from strains. Results show that the stresses can be evaluated from in-plane strains even if the Poisson's ratio exhibits time- and temperature-dependence more accurate than the developed method.

Keywords Viscoelasticity · Plane stress · Numerical Laplace transformation · Pseudoelasticity

6.1 Introduction

Not only in-plane strains but a through-thickness strain exists in an object under plane stress condition even if a through-thickness stress is vanishingly small. In the case of an elastic body, in-plane stresses can be computed from in-plane strains because a through-thickness strain is obtained from in-plane strains through a relation with a constant Poisson's ratio. A similar situation can be seen in the case of a plastic body which exhibits incompressibility. In the case of a viscoelastic body, on the other hand, a through-thickness strain cannot be obtained from in-plane strains because the Poisson's ratio depends on time and temperature. Therefore, it is difficult to evaluate stresses from measured values of in-plane strains in a viscoelastic body under plane stress condition.

Conventionally, the Poisson's ratio of a viscoelastic body is often treated as a constant [1, 2]. In this case, in-plane stresses can be obtained from in-plane strains as with elastic and plastic bodies. However, the Poisson's ratio actually exhibits time and temperature-dependence [3–7], and thus, the assumption of a constant Poisson's ratio can induce an error in stress analysis.

The stress-strain relation of a viscoelastic body is represented by a higher-order differential equation. The constitutive equation of a linear viscoelastic material is represented in the same form as an elastic body in Laplace domain. This relation is called the correspondence law. Since the relation between a through-thickness strain and in-plane strains is expressed in Laplace domain, Poisson's ratio can be treated as time- and temperature-dependence. The authors have proposed a method for evaluating stresses from measured in-plane strains of the viscoelastic body under a plane stress condition using numerical Laplace transformation [8]. By using this method, stresses can be evaluated from in-plane strains even if Poisson's ratio exhibits time- and temperature-dependence. However, the accuracy of calculation by the method for computing stresses using numerical Laplace transformation depends on the constitutive equation in Laplace domain. Therefore, it is necessary that stresses can be evaluated more accurately.

S. Taguchi (✉) · K. Takeo · S. Yoneyama
Department of Mechanical Engineering, Aoyama Gakuin University, Sagamihara-shi, Kanagawa, Japan
e-mail: c5619123@aoyama.jp; takeo@me.aoyama.ac.jp; yoneyama@me.aoyama.ac.jp

M.-T. Lin et al. (eds.), *Advancement of Optical Methods & Digital Image Correlation in Experimental Mechanics*,
Conference Proceedings of the Society for Experimental Mechanics Series, https://doi.org/10.1007/978-3-030-59773-3_6

6.2 Objective

This study proposes a method for evaluating stresses from measured in-plane strains of the viscoelastic body under a plane stress condition. A through-thickness strain is obtained from in-plane strains based on pseudoelasticity in the linear viscoelastic theory. That is, the in-plane strains as well as the through-thickness strain is computed in a Laplace domain. The Laplace transformation is performed numerically using a method developed by the authors [8]. The stresses are computed from the through-thickness strains and in-plane strains and by numerical integration. The effectiveness of the proposed method is demonstrated by applying it to viscoelastic stress analysis.

6.3 Constitutive Equation of Viscoelasticity

The three-dimensional constitutive equations of viscoelasticity are expressed by [9].

$$\sigma_{kk}(t) = 3K(t)\varepsilon_{kk}(0) + 3\int_0^t K(t-\tau)\frac{\partial \varepsilon_{kk}(\tau)}{\partial \tau}d\tau \tag{6.1}$$

$$s_{ij}(t) = 2G(t)e_{ij}(0) + 2\int_0^t G(t-\tau)\frac{\partial e_{ij}(\tau)}{\partial \tau}d\tau \tag{6.2}$$

where $\sigma_{kk}(t)$ is the sum of normal stress components, $s_{ij}(t)$ is the deviatoric stress, $e_{ij}(t)$ is the deviatoric strain, and $K(t)$ is the relaxation bulk modulus and $G(t)$ are relaxation shear modulus. It can be recognized from Eqs. (6.1) and (6.2) that the through-thickness strain $\varepsilon_{zz}(t)$ is required for computing stresses from strains even if an object is under the plane stress condition. Because the Poisson's ratio of a viscoelastic material usually depends on time and temperature, it is difficult to obtain through-thickness strains. In the present study, the through-thickness strains are obtained based on pseudoelasticity in the linear viscoelastic theory. This is the property that the constitutive equation of a linear viscoelastic material is represented in the same form as an elastic body in Laplace domain. Therefore, the through-thickness strain in Laplace domain is expressed as

$$\tilde{\varepsilon}_{zz}(s) = -\frac{s\tilde{\upsilon}(s)}{1 - s\tilde{\upsilon}(s)}\left(\tilde{\varepsilon}_{xx}(s) + \tilde{\varepsilon}_{yy}(s)\right) \tag{6.3}$$

where $\tilde{\varepsilon}_{zz}(s)$, $\tilde{\upsilon}(s)$, $\tilde{\varepsilon}_{xx}(s)$,and $\tilde{\varepsilon}_{yy}(s)$ are the through-thickness strain, the Poisson's ratio, and the normal strains in the Laplace domain, respectively. The through-thickness strain $\varepsilon_{zz}(t)$ is obtained from $\tilde{\varepsilon}_{zz}(s)$ by Laplace inverse transformation. Then, the stresses can be computed from obtained values and in-plane strains by the numerical integration using Eqs. (6.1) and (6.2).

6.4 Numerical Laplace Transformation Using FFT

Taking the variable s in the Laplace transformation and the inverse transform equations as $s = \gamma + i\omega$ and considering that $f(t) = 0$ at $t < 0$, the Laplace transformation can be expressed as [9]

$$F(\omega) = \int_{-\infty}^{\infty} f(t)\cdot e^{-\gamma t}\cdot e^{-i\omega t}dt \tag{6.4}$$

$$f(t) = \frac{e^{\gamma t}}{2\pi}\int_{\gamma - i\infty}^{\gamma + i\infty} F(\omega)\cdot e^{i\omega t}d\omega \tag{6.5}$$

where $F(\omega)$ is the Laplace transformation of the function $f(t)$. It is observed that the Laplace transformation and the inverse transformation can be represented in the form of the Fourier transformation and the inverse transformation. When the infinite sum of integrals is truncated by the $N - 1$ term and the number of data is N, Eqs. (6.4) and (6.5) can be expressed as

$$F(n\Delta\omega) = \Delta t \sum_{k=0}^{N-1} f(k\Delta t) \cdot e^{-\gamma k \Delta t} / e^{-2\pi i k n / N} \tag{6.6}$$

$$f(k\Delta t) = \frac{e^{\gamma k \Delta t}}{2\pi} \Delta\omega \sum_{k=0}^{N/2-1} F(n\Delta\omega) \cdot e^{2\pi i k n / N} \tag{6.7}$$

where $\Delta\omega$, Δt, n, and ω are the angular frequency interval, the time interval, the data number, and the angular frequency, respectively. It is possible to utilize fast Fourier transformation (FFT) for the above equation, and the computation can be performed by setting N to a power of 2. When computing using FFT, the input function is regarded as one cycle. Assuming that this period is $N\Delta t$, the appropriate value of γ is $\gamma = 4/(N\Delta t)$~$5/(N\Delta t)$ [9]. However, the appropriate values of the numerical Laplace transformation cannot be obtained because the results are affected by the computational nature of FFT.

6.5 Numerical Laplace Transformation Considering the Nature of FFT

This section shows a computation method that is not affected by the nature of FFT.

The results of the numerical Laplace transformation by FFT are effective in the range of $0 < \omega < \omega_s/2$ where ω_s is the sampling angular frequency. This is caused by the FFT nature where the discrete data is used. The results are obtained in the range of $0 < \omega < \omega_s$ by doubling the number of the input strain data. Since the number of the input data is doubled, ω_s is also doubled but ω is constant. Therefore, the results in the range of $0 < \omega < \omega_s$ is obtained appropriate values.

When the Laplace inverse transformation is computed using the effective values in the range of $0 < \omega < \omega_s/2$, appropriate values cannot be obtained. It is considered that transformation solutions of the input used for the Laplace inverse transformation using FFT need the complex conjugate. Therefore, transformation solutions of conjugate complex data are obtained by doubling the number of data. Then, the additional data are plotted symmetrically with the value at $\omega_s/2$ as the boundary. Since the number of the input data is doubled, $\omega_s/2$ is also doubled but $\Delta\omega$ is constant.

6.6 Computing Stresses Using Proposed Method

In order to show the effectiveness of the numerical Laplace transformation and the inverse transformation methods proposed in previous section, the through-thickness strains are calculated from in-plane strains. Also, the stresses are calculated from obtained values and in-plane strains using constitutive equation of the viscoelastic body.

$K(t)$, $G(t)$, and $\upsilon(t)$ are expressed as by using a Prony series as follows expressed as [10]

$$K(t) = K_0 + \sum_{i=1}^{N} K_i e^{-t/\rho_i} \tag{6.9}$$

$$G(t) = G_0 + \sum_{i=1}^{N} G_i e^{-t/\rho_i} \tag{6.10}$$

$$\upsilon(t) = \upsilon_0 - \sum_{i=1}^{N} \upsilon_i e^{-t/\rho_i} \tag{6.11}$$

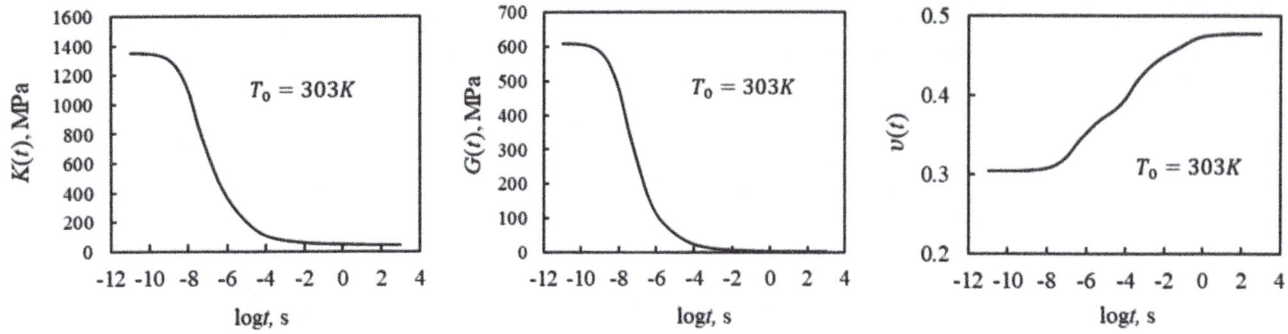

(a) Relaxation bulk modulus (b) Relaxation shear modulus (c) Poisson's ratio

Fig. 6.1 Master curves of each material property

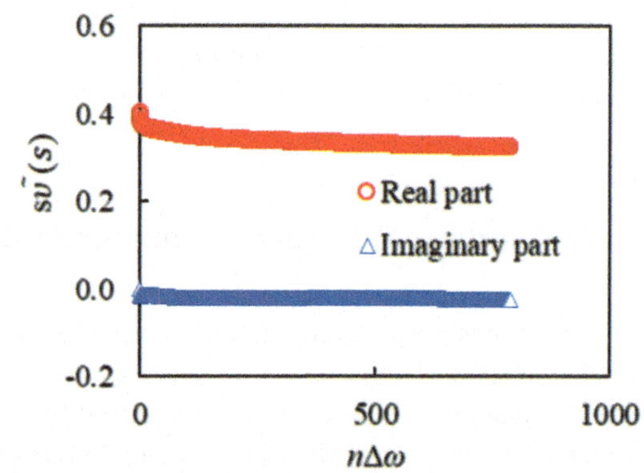

Fig. 6.2 Poisson's ratio in the Laplace domain ($T = 273$ K, $s = \gamma + i\omega$, $\gamma = 0.142$, $\Delta\omega = 0.767$ rad/s)

where k_0, k_i, G_0, G_i, υ_0, and υ_i are the coefficients of the Prony series, N is the number of terms in the series, and ρ_i is the relaxation time.

Figure 6.1 shows master curves of the material properties at the reference temperature $T_0 = 303$. The material is a soft epoxy resin (Epikote 871).

$\tilde{\varepsilon}_{zz}(s)$ is computed using Eq. (6.3). In the equation, $s\tilde{\upsilon}(s)$ can be obtained by calculating the Laplace transformation of $\upsilon(t)$ and multiplying s on both sides of the equation as follows:

$$s\tilde{\upsilon}(s) = \upsilon_0 - \sum_{i=1}^{M} \frac{s\rho_i \upsilon_i}{s\rho_i + 1} \tag{6.12}$$

Poisson's ratio in the Laplace domain is obtained by Eq. (6.12). Since the Fourier transformation is used, taking into considering that the variable $s = \gamma + i\omega$, $s\tilde{\upsilon}(s)$ is shown in Fig. 6.2.

Figure 6.2 shows the Poisson's ratio in the Laplace domain. The temperature is $T = 273$ K. $\tilde{\varepsilon}_{zz}(s)$ is calculated by using the above property. The input strains is shown in Fig. 6.3.

Δt is $\Delta t = 0.008$, and the number of original input data is 1024. According to the proposed method, the number of original data is increased to 2048 for calculation. Each parameter is $\gamma = 0142$ and $\Delta\omega = 0.767$ rad/s. Figure 6.4 shows the calculation results of the through-thickness strains calculated by Eqs. (6.3) and (6.7) using the input values obtained by the proposed method.

Calculation results are compared with the values obtained by 3D FEM. There are 1024 calculation results. Calculation results are generally consistent. Next, the stresses are computed from the obtained values, in-plane strains, and material properties by the numerical integration using Eqs. (6.1) and (6.2). Figure 6.5 shows the calculation results of normal stress components.

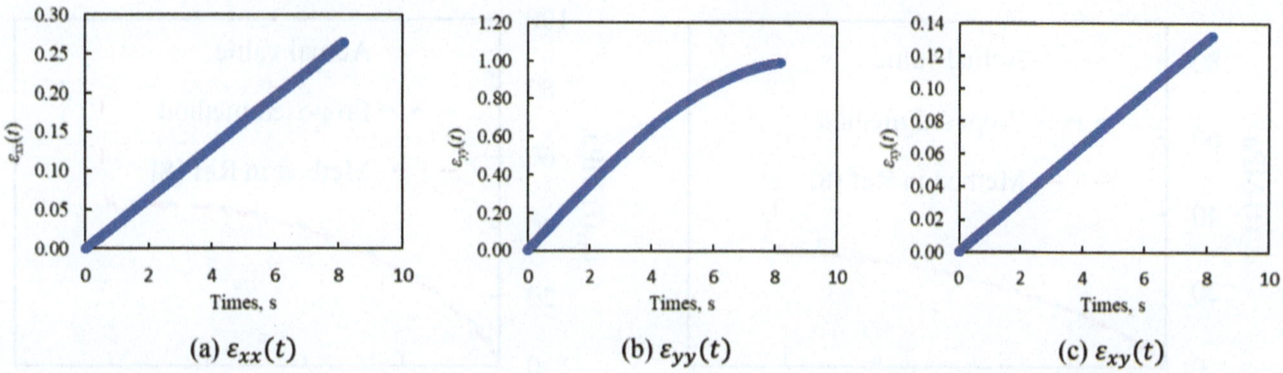

Fig. 6.3 Input strains

Fig. 6.4 Calculation results of through-thickness strains

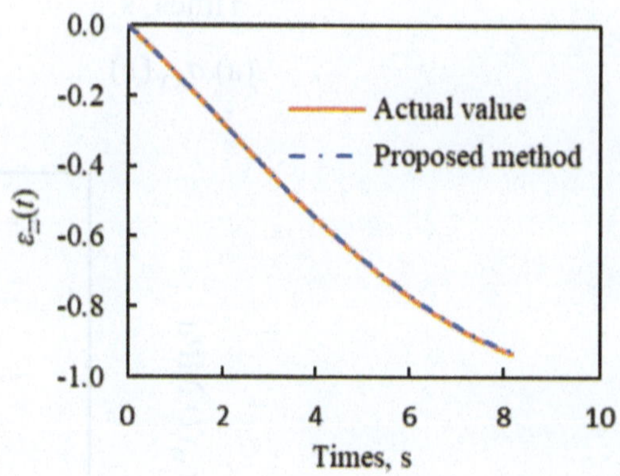

Each result is compared with the values obtained by 3D FEM and computed from the constitutive equations in Laplace domain using numerical Laplace transformation. The calculation results are generally more consistent than the values computed by the numerical Laplace transformation especially near the end of the data. It is shown that in-plane stresses can be obtained by the proposed method.

6.7 Conclusions

This study proposes a method for computing stresses from measured values of in-plane strains in viscoelastic body under plane stress condition. Since the relation between a through-thickness strain and in-plane strains is expressed in Laplace domain, Poisson's ratio can be treated as time- and temperature-dependence. The stresses are computed from the through-thickness strains and in-plane strains by numerical integration. The effectiveness of the proposed method is demonstrated by computing stresses from strains. The values obtained by the proposed method are compared with the values obtained by 3D FEM and computed from the constitutive equations in Laplace domain using numerical Laplace transformation. The calculation results are generally more consistent than the values computed by the numerical Laplace transformation especially near the end of the data. It is shown that in-plane stresses can be obtained by proposed method. In addition, stresses can be calculated by taking Poisson's ratio into consideration of the time dependency by using the proposed method. In the future, it is necessary to compute stresses from various strains and to examine effectiveness.

Acknowledgments This work is supported by JSPS KAKENHI Grant Number JP18K03845.

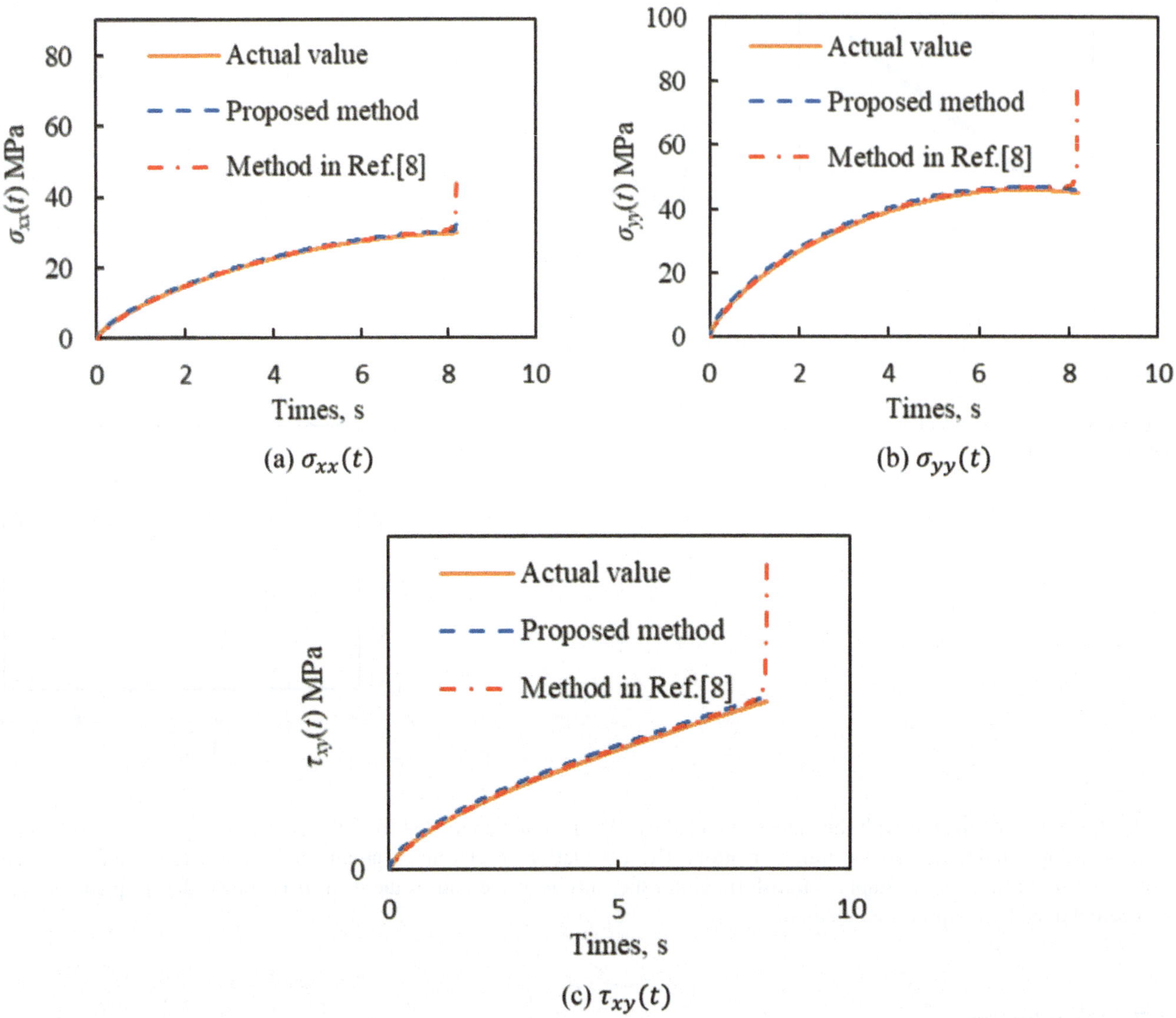

Fig. 6.5 Calculation results of stresses in each direction

References

1. Yoneyama, S., Ogawa, K., Misawa, A., Takashi, M.: Evaluation of time-dependent fracture mechanics parameters of a moving crack in a viscoelastic strip. JSME Int. J. Ser. A. **42**(44), 624–630 (1999)
2. Kaku, K., Arai, M., Fukuda, T.: Evaluation of thermo-viscoelastic property of CFRP laminate based on homogenization theory. Acta Mech. **214**(1), 111–121 (1996)
3. Tschoegl, N.W., Knauss, W., Emri, G.: Poisson's ratio in linear viscoelasticity a critical view. Mech. Time-Depend Mater. **6**(1), 3–51 (2002)
4. Yee, A.F., Takemori, M.T.: A new technique measuring volume and shear relaxations in solid polymers: dynamic Poisson's ratio. ACS Polym. Preprints. **20**(1), 758–761 (1979)
5. Giovagnoni, M.: On the direct measurement of the dynamic Poisson's ratio. Mech. Mater. **17**(1), 33–46 (1994)
6. Priz, T.: Measurement methods of complex Poisson's ratio of viscoelastic materials. Appl. Acoust. **60**(3), 279–292 (2000)
7. Emri, I., Prodan, T.: A measuring system for bulk and shear characterization of polymers. Exp. Mech. **46**(4), 429–439 (2006)
8. Taguchi, S., Takeo, K., Yoneyama, S.: Computing stresses from measured in-plane strains in viscoelastic body under plane stress condition. In: Proceedings of the 14th International Symposium on Advanced Science and Technology in Experimental Mechanics, Paper FL387, pp. 1–6 (2019)
9. Inoue, H., Kamibayashi, M., Kishimoto, K., Shibuya, T., Koizumi, T.: Numerical Laplace transformation and inversion using the Fast Fourier Transform. JSME Int. J., Ser. 1. **35**(3), 319–324 (1992)
10. Gutierrez-Lemini, D.: Engineering Viscoelasticity, pp. 23–30. Springer, New York (2014)

Chapter 7
Evaluation of the Influence of Water Absorptivity on the Properties of CFRP Cylinder Materials by SHPB Impact Compression Test

J. Liu, K. Takeo, and S. Yoneyama

Abstract In this study, dynamic compressive behaviors of resin-based CFRP were measured by using split Hopkinson pressure bar tests to consider the effect of water absorption on dynamic mechanical properties. The split Hopkinson pressure bar test method (SHPB method) is used to provide high strain rate. In order to satisfy the experimental equipment needs of SHPB method, cylindrical CFRP material was used for impact compression test.

Similar sample with different water absorption rates after several days of continuous water absorption under hygrothermal conditions were used in some of the experiments. This fiber interlayer and resin matrix water absorption will pose a great influence on the strength and strain rate of the material and even the load rate dependency under the compression of the vertical fiber direction. The process of destruction was observed using the experimental mechanics, and the experimental data was evaluated.

Keywords SHPB impact compression tester · Water absorption rates · Strength-strain rate dependence

7.1 Background

Carbon fiber composites such as Carbon Fiber Reinforced Plastic (CFRP) have higher specific strength, rigidity, and specific gravity compared with general metal and metal composites. Therefore, these materials are currently commonly used in the manufacture of various components of aircraft, such as the wing of the aircraft, the hatch and the cabin casing, etc. Materials used in aircraft are required to have impact resistance and vibration resistance in high humidity environments. CFRP material expands and deforms, generating residual stress and changing the material properties in a high-temperature and humidity environment [1].

FRP materials are often manufactured by Resin Transfer Molding, Sheet Molding Compound, and Autoclave method. However, these methods create small hollow pores between the composite plates during the pressure forming process. The size, shape, and amount of pores have a great effect on mechanical properties such as strength. It is necessary to understand the effect of water absorption on the mechanical properties due to the softening of fiber matrix in the high-temperature and humidity environment [2].

In previous studies, Hironaka and Yamamoto et al. clarified that CFRP plate absorbed for a long time with 373 K of distilled water had improved surface softening and Charpy impact value [3, 4]. Meng et al. conducted a three-point bending test of CFRP laminates in a high-temperature and humidity environment to investigate the effect of pores generated during the forming process on the water absorption of the laminates. In the results, the water absorption has obvious influence on the material strength decline [5]. However, the effect of moisture absorption of resin matrix under dynamic high strain rate was not obtained. The purpose of this study is to analyze the effect of water absorption on the dynamic compression characteristics of CFRP.

J. Liu (✉) · K. Takeo · S. Yoneyama (✉)
Department of Mechanical Engineering, Aoyama Gakuin University, Sagamihara-city, Kanagawa, Japan
e-mail: c5619133@aoyama.jp; takeo@me.aoyama.ac.jp; yoneyama@me.aoyama.ac.jp

M.-T. Lin et al. (eds.), *Advancement of Optical Methods & Digital Image Correlation in Experimental Mechanics*, Conference Proceedings of the Society for Experimental Mechanics Series, https://doi.org/10.1007/978-3-030-59773-3_7

7.2 Specimen

First, the water absorption ratio was measured to clarify saturated water absorption capacity of CFRP material. Epoxy resin was used for the matrix of CFRP in this test. As shown in the Fig. 7.1, CFRP had width of 12 mm, length of 20 mm, and thickness of 2 mm. The fiber directions were 0 and 90 degrees, and the laminate structure has a total of 15 layers. The content of resin and density of CFRP were $40 \pm 2W_t\%$ and 1.5 g/cm^3, respectively. Percentage of volatile components is less than $2.0W_t\%$.

The specimen used in the impact compression test is shown in Fig. 7.2. The CFRP was 12 mm in inner diameter, 14 mm in outer diameter, and 7 mm in thickness. The fiber direction was 90 degrees' perpendicular to the compression direction.

7.3 Experiment

First, drying treatment was conducted for the specimen in a drying furnace at a constant temperature of 343 K. The processing time is set to 24 h so as to dry specimen completely. Nine specimen were put in distilled water at 323 K and 298 K (normal temperature), respectively, and specimen mass was measured from 2 h to 2 weeks after the water absorption experiment. The calculation formula of water absorption is expressed as follows:

$$C = (M_t - M_0) / M_0 \times 100\%$$

Where C, M_0, and M_t are mass change rate, mass before water absorption, and mass after water absorption.

In terms of impact compression test, the split Hopkinson pressure bar test method (SHPB method) is used to provide high strain rate. In order to satisfy the experimental equipment needs of SHPB method, cylindrical CFRP material was used for impact compression test. The input bar, output bar, and striker of the SHPB testing machine were all solid bars with a diameter of 16 mm made of SUJ2 high-carbon chromium bearing steel. The input bar was 2500 mm in length, the output bar was 1300 mm in length, and striker was 1000 mm in length, respectively. As shown in Fig. 7.3, according to the one-dimensional stress wave theory, the dynamic stress-strain curve and the dependence relationship between compression

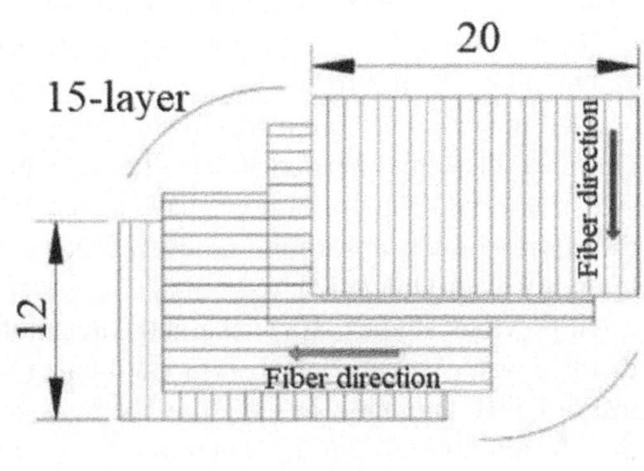

Fig. 7.1 Water absorption test specimen

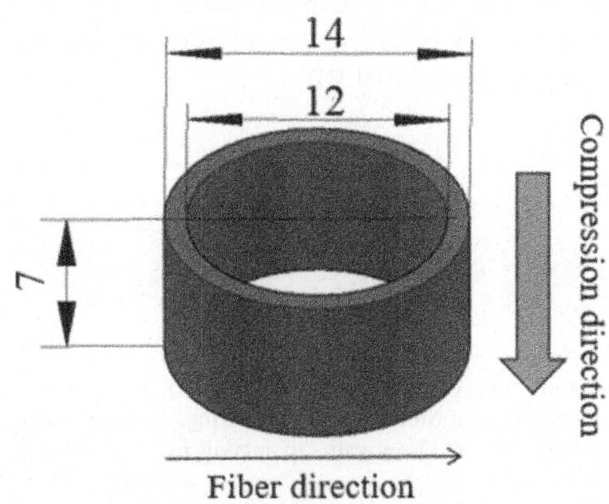

Fig. 7.2 Impact test specimen

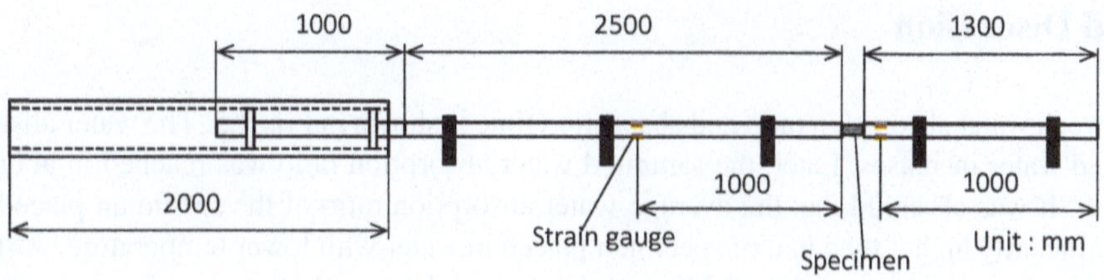

Fig. 7.3 Hopkinson bar method

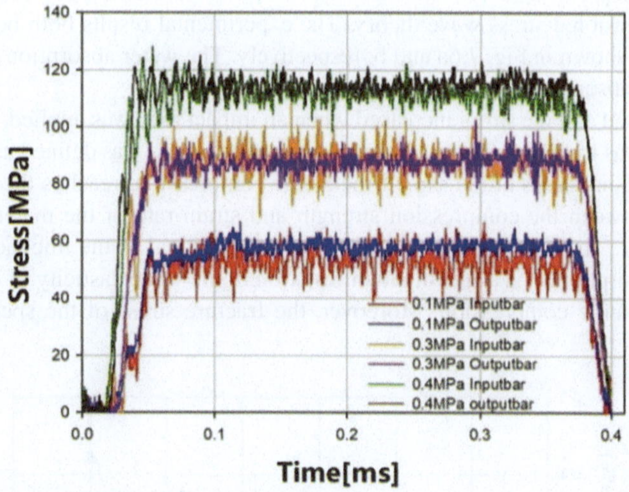

Fig. 7.4 Input and output bar side stress after calibration (0.1~0.4 MPa)

strength and strain rate can be measured. Stress σ, strain rate, and strain can be expressed by using an incident, reflected, and transmitted strain waves ε_i, ε_r, and ε_t.

$$\sigma = \frac{EA}{2A_S}(\varepsilon_i + \varepsilon_r + \varepsilon_t)$$

$$\dot{\varepsilon} = \frac{E}{\rho C_0 l_S}(\varepsilon_i - \varepsilon_r - \varepsilon_t)$$

$$\varepsilon = \int_0^t \dot{\varepsilon} dt$$

where E, ρ, C_0, A, A_s, and l_s are Young's modulus of the input and output bars, the density of bars, the velocity of elastic wave, area of bars, area of the specimen, and gauge length of the specimen.

The accuracy of the SHPB testing machine used in this study was verified. The testing machine was calibrated under air pressure 0.1~0.4 MPa. The input rod and output rod of the testing machine were directly fitted together without the specimen. The stress wave after striker collision was measured from strain gauges bonded to input and output bars. Based on the one-dimensional stress wave theory, the axial stress at the joint parts between the input and output ends was calculated. Figure 7.4 is shown the axial stresses. Furthermore, since noise occurred due to the deflection of the outer end of the input and output bars, the deflection of the end was eliminated by using a jig.

7.4 Results and Discussion

The relationship between water absorption ratio and absorption time is shown in Fig. 7.5. The water absorption ratio increased as the time in distilled water increases. Later, the saturated water absorption ratio was reached in about 120 h regardless of the water temperature. It was observed that the average water absorption ratio of the specimen placed in water with higher temperature was significantly higher than that of specimen placed in water with lower temperature. As the water temperature increased 25 °C, the increase rate increased by 0.2%, and the saturated water absorption was nearly doubled.

Next, the dynamic compressive behavior of CFRP both before and after water absorption was investigated. The striker was launched under the air compressive of 0.1~0.4 MPa. The impact stress-strain curve and strain-strain rate curve were obtained through the one-dimensional stress wave theory. The experimental results both before and after absorption under the air pressure of 0.4 MPa are shown in Fig. 7.6a and b, respectively. The water absorption test specimen absorbed distilled water of 343 K until the water absorption reached 0.4%.

The specimen was compressed and the stress increased when an impact load was applied. Afterwards, the stress gradually decreases when the specimen was fractured. The maximum compressive stress was defined as the compressive strength of the CFRP. Similarly, the strain rate increases at the same time as the compression, reaches the maximum value, and gradually decreases. The relationship between the compression strength and strain rate of the material both before and after water absorption is shown in Fig. 7.7. The compressive stresses were proportional to the thickness for the strain rate regardless both before and after water absorption. It is a specific manifestation of the viscoelasticity of the resin matrix of the interlayer of composite fibers under dynamic compression. Moreover, the fracture stress of the specimen that absorbed water was

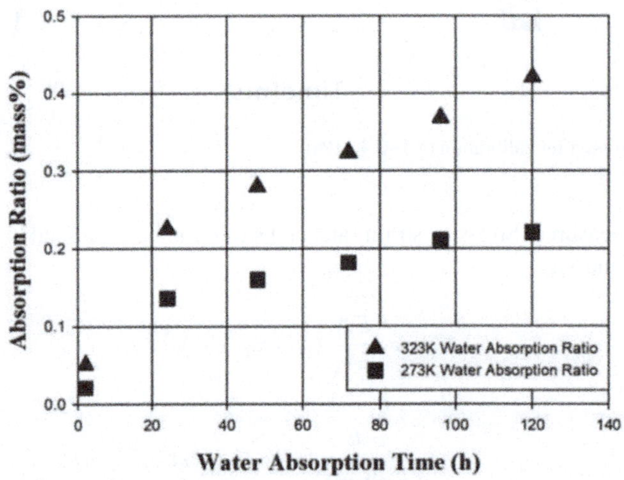

Fig. 7.5 Changes in water absorption ratio against absorption time

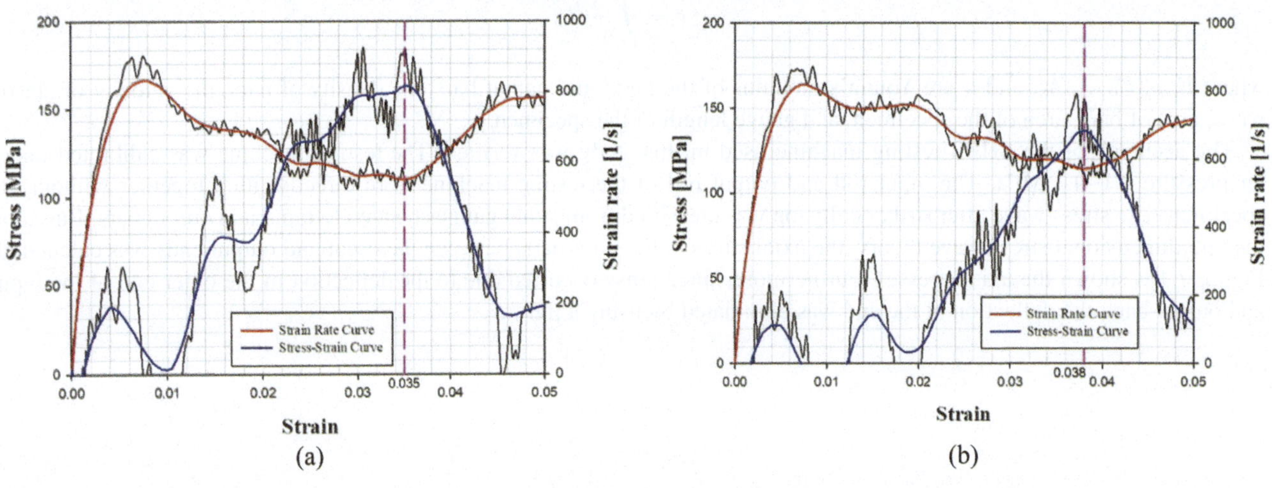

Fig. 7.6 Dynamic characteristics of specimen before (**a**) and after (**b**) water absorption (0.4 MPa)

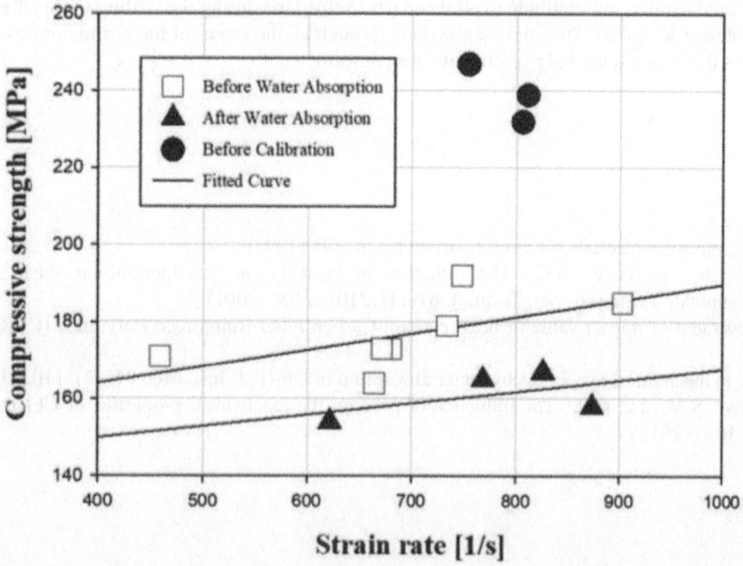

Fig. 7.7 Description of image

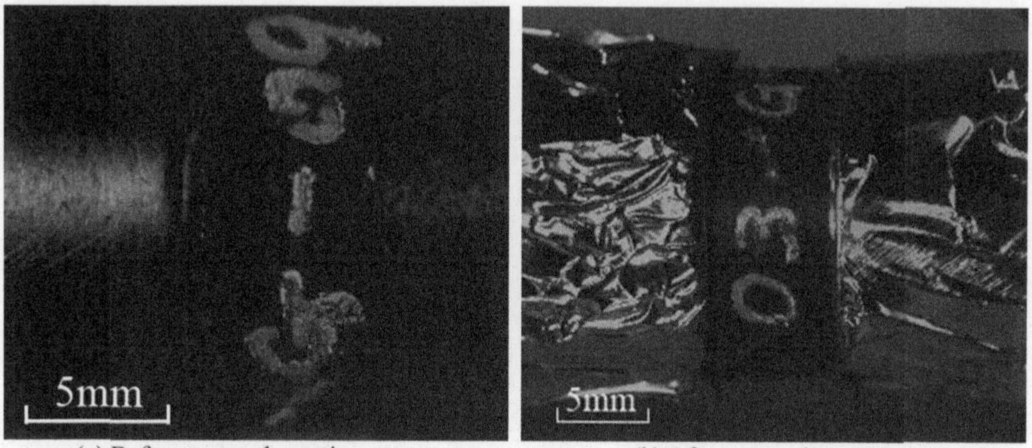

(a) Before water absorption (b) After water absorption

Fig. 7.8 Destruction of specimen

smaller than that before water absorption, indicating that the compressive strength decreased due to the softening of resin and fiber.

Finally, fracture process of specimen was observed by using high-speed camera. Figure 7.8a and b are shown fracture specimen both before and after water absorption, respectively. The specimen fractured in the 45° direction regardless of the water absorption ratio. Therefore, the water absorption has no significant effect on the fracture behavior and fracture mode of CFRP.

7.5 Conclusion

In this study, the effects of water absorption ratio on strain rate and compressive strength were investigated experimentally by using the SHPB test method at a strain rate of 200~1000s^{-1}. The compressive strength exhibited strain rate dependence regardless of the water absorption ratio of specimen. In addition, the compressive strength of the specimen decreased due to water absorption, and the strain rate dependence was reduced. Finally, the fracture of CFRP both before and after water absorption was observed by using the high-speed camera.

Acknowledgments I would like to express my gratitude to all those who helped me during the writing of this thesis. Special thanks to Professor S. Yoneyama and Assistant Professor K. Takeo. They have walked me through all the stages of the writing of this thesis. Without their consistent and illuminating instruction, this thesis could not have reached its present form.

References

1. Hoskin, B.C., Baker, A.A: Composite materials for aircraft structures. p. 144 (1992)
2. Costa, M.L., Dealmeida, S.F.M., Rezende, M.C.: The influence of porosity on the interlaminar shear strength of carbon/epoxy and carbon/bismaleimide fabric laminates. Compos. Sci. Technol. **61**(14), 2101–2108 (2001)
3. Yamamoto, T., et al.: Improvement of impact value of heat resistant Carbon Fiber Reinforced Polyimide (CFRP) by water absorption. J. Inst. Met. **74**(4), 274–278 (2010)
4. Hironaka, N., et al.: Change in mechanical properties by water absorption of CFRP. J. Inst. Met. **71**(12), 1102–1106 (2007)
5. Meng, M., Rizvi, M.J., Grove, S.M., Le, H.R.: The influence of pore on the mechanical properties of CFRP on hygrothermal environment. Compos. Struct. **133**, 1024–1035 (2015)

Chapter 8
Speckling and Testing with DIC at Microscales

Kevin B. Connolly and W. Carter Ralph

Abstract The full-scale performance of modern materials, such as composites and additively manufactured parts, is dependent on their characteristic substructure. An understanding of the material interactions within these repeating substructures can improve the understanding of the material's bulk properties. Taking measurements within these substructures is made difficult by their small size and potentially non-homogenous nature. There are some techniques that already exist for making small-scale strain measurements, but these techniques can require extensive specimen preparation and expensive specialty equipment and can place significant limitations on the specimen and test design. Southern Research is working to develop a method to measure strain in these substructures using Digital Image Correlation (DIC). This approach relies on a well-designed experimental setup to apply thermal loads and observe the resulting strains. The project design has moved through a number of iterations to address the various difficulties in capturing high magnification, high-resolution images with the correct quality for use with DIC.

Keywords Strain measurement · Digital image correlation · Thermal strain · Optical microscopy · Surface preparation

8.1 Introduction

Complex materials are becoming more common in both research and commercial sectors. Examples include the structures designed in modern micro-processors, variations in material properties associated with 3D printing methods, and material mismatches inherent to composite materials. The variations in these materials are highly localized making detailed analysis of the structures difficult. The details of these substructures can be useful when determining the performance of the full component and improving the design. Traditional strain measurements do not have a high-enough resolution to distinguish between different components of the structure.

This project is a continuation of previous work [1] to investigate the possibility of using Digital Image Correlation (DIC) as a method for measuring strain on small structures. DIC uses a series of images to calculate the displacement and strain fields on a surface by tracking common features in the image and comparing them back to a reference image. Methods have existed for measuring small strain fields, but these techniques tend to require difficult specimen preparation, require expensive equipment, and place restrictions on the specimen size or test method. The versatility of DIC is the primary reason it is being investigated for this work. DIC does not place specific limitations on patterning designs, so specimen preparation tends to be inexpensive and quick. Capturing images for DIC processing can be done with any camera system. Finally, DIC can be performed in most environments, so there are fewer limitations on specimen size and testing fixtures.

The purpose of this work is to measure the stiffness mismatch in the small scale structures previously mentioned. In the current test setup, the specimen is thermally loaded to create strain in the material. Two fields of view have been considered in this work, approximately 1.2 mm × 1.0 mm and 0.4 mm × 0.3 mm field of view. These fields of view were chosen so that components of microprocessors could be visible and distinct. As this project has progressed, technical problems have been addressed. This paper will report on the two issues that have given the most difficulty, specimen patterning, and noise in the strain measurements.

K. B. Connolly (✉) · W. C. Ralph
Southern Research Institute, Birmingham, AL, USA
e-mail: kconnolly@southernresearch.org; cralph@southernresearch.org

M.-T. Lin et al. (eds.), *Advancement of Optical Methods & Digital Image Correlation in Experimental Mechanics*,
Conference Proceedings of the Society for Experimental Mechanics Series, https://doi.org/10.1007/978-3-030-59773-3_8

8.2 Apparatus

The project apparatus consists of a microscope, specimen stage, heater, and super structure to hold components in alignment (see Fig. 8.1). The design of the test setup is covered by patent 10,379,333. The microscope currently being used is a 5 megapixel camera attached to telocentric lens. The objective lens on this system is interchangeable and in most cases results in a working distance of approximately 40 mm. The specimen stage consists of a two-axis adjustable base for in-plane motion, insulated supports, and a low thermal expansion stage that supports the specimen and heater. The low thermal expansion stage includes a hole to allow a clear optical path to the specimen. Holes in the stage also allow room for a thermocouple to lay under the specimen and measure temperature from the same surface that is being viewed for DIC measurements. The microscope lens is also instrumented with a thermocouple. The apparatus uses a simple resistance heater with holes to ensure that the air flow is sufficient to prevent air stagnation in the optical path to the microscope. The primary purpose of the superstructure is to reduce vibrations into system and ensure that the microscope is held perpendicular to the specimen stage.

Testing is conducted by taking image sets at predetermined temperatures. As the temperature of the specimen reaches the appointed mark, the camera is triggered, and the temperature of the specimen and lens is recorded. After heating up to the desired temperature, the apparatus is allowed to cool naturally and additional images are taken for analysis. For most test runs, the specimen was heated to 170 °C before being allowed to cool. Once the test is concluded, the images sets are imported into DIC software for analysis.

8.3 Specimen Preparation

DIC requires high contrast and sufficiently small features to track displacement of the surface properly. Creating an acceptable pattern for the desired fields of view can be difficult. The optimum size for contrast features is 3–5 pixels [2]. For the 1.2 mm by 1.0 mm field of view, this would require features about 2.5 μm in size. For the 0.4 mm by 0.3 mm field of view, this would require features about 0.8 μm in size. The most successful DIC pattern for this project has relied on the natural microstructure of materials. This worked reasonably well for aluminum and certain microprocessors in mid process, shown in Figs. 8.2 and 8.3, respectively. Unfortunately, most materials of interest do not produce an acceptable pattern when prepared for microscopy.

The standard method for applying a DIC pattern is to sputter paint using a spray can or an airbrush for smaller specimens. Standard black and white airbrush paint did not produce enough contrast at high magnification. It was found that tempera

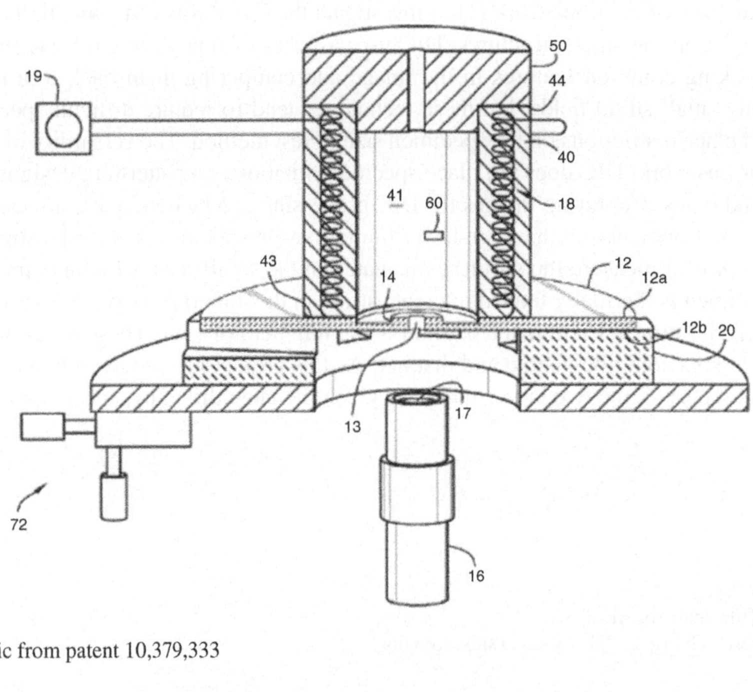

Fig. 8.1 Apparatus schematic from patent 10,379,333

Fig. 8.2 As-polished surface features on an aluminum sample, 1.2 mm by 1.0 mm field of view

Fig. 8.3 As-manufactured surface features on a second level copper pad

paints could be used with an airbrush and produced acceptable contrast for tracking (Fig. 8.4). Even with sufficient contrast, the feature size with airbrushed paints was too large for the fields of view being attempted.

Another technique used for creating a pattern is the use of micro stamp SEM pads. These stamp pads are intended to create a topography for performing DIC in an SEM but can be used on reflective substrates to create an optical pattern. These stamps are designed to have a 1 μm feature size. Unfortunately, the consistently of the pattern was poor (Fig. 8.5). Strips of useful pattern are separated by less distinct regions. Even at the smallest field of view, the distinct areas were not large enough to cover the image. Due to the variability in definition, this technique is also not suitable for covering a particular area of interest on a specimen.

Fig. 8.4 Aluminum painted with airbrushed Tempra paint, 1.2 mm by 1.0 mm field of view

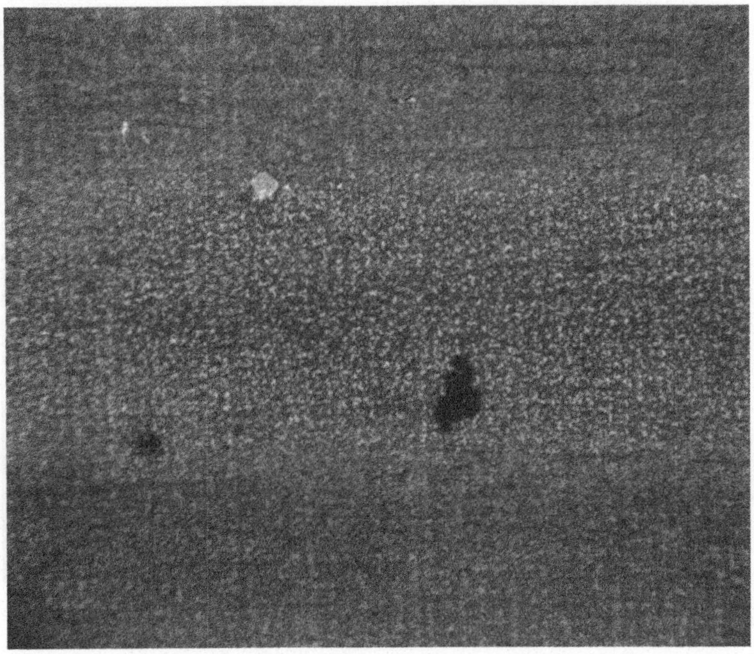

Fig. 8.5 Aluminum with SEM stamp pad, 0.4 mm by 0.3 mm field of view

The most recent technique being explored is to use carbon powder, size 0.4–12 μm, to create a DIC speckle. This powder is deposited on a freshly painted specimen surface so that it adheres to the surface as the paint dries. Similar to the air brush, there were difficulties with getting proper contrast and feature sizing (Fig. 8.6). However, this technique created a more repeatable pattern than efforts using an airbrush.

Fig. 8.6 Aluminum painted white with carbon powder deposited, 1.2 mm by 1.0 mm field of view

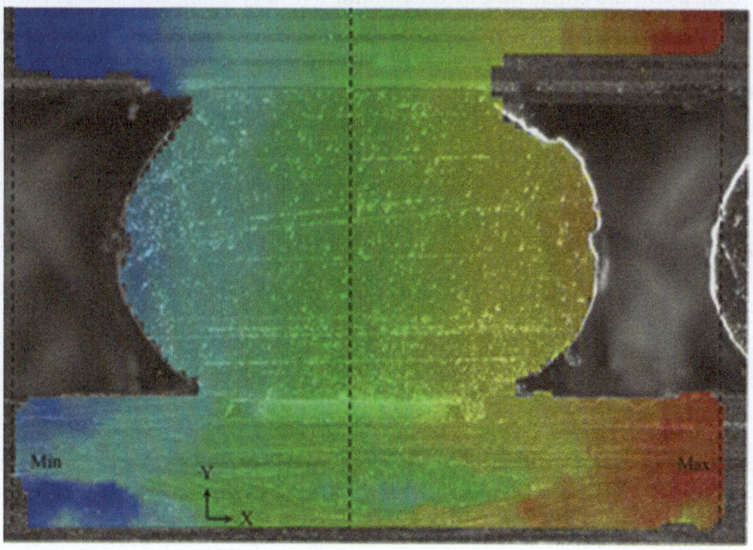

Fig. 8.7 Relative X displacement of second-level interconnect at 170 °C

8.4 DIC Results

Early results illustrated that DIC could track the movement of non-homogeneous structures. Figure 8.7 shows the displacement field of a second-level interconnect in a microprocessor along with a slight distortion in the strain field of the substrate where the material is not uniform. Trying to validate the test measurements proved more difficult. For validation, DIC was used to measure the thermal expansion of a simple aluminum specimen. Test results were fairly accurate at measuring the expansion of the aluminum, but noise was a significant problem. The noise in the system was higher than the strain signal until the temperature of the specimen was above 70 °C. Even at the maximum applied temperature, 170 °C, the noise was still equal to about one quarter of the total signal.

The noise seen in tests also increased with temperature. By taking a quick sequence of images at temperature, there was no evidence of temporal variations in the images. Changes in the stain over a short time could be associated with diffraction of air due to heat, but were not present. Another cause for the increasing noise with temperature could be the temperature change of the microscope lens. It is possible that the increase in noise was due to the slight change in focus caused by changes in the lens length. Trying to insulate the lens does not affect the changes in the lens temperature. It is likely the clear path between the specimen and lens is a cause for the heating. One possible technique to address this problem could be active cooling of the lens using ice water to maintain a consistent temperature throughout the test.

In addition to thermal causes for noise, there may also be problems stemming from the flatness of the acquired images. As the temperature increases, the specimen tends to move slightly as it expands. Exploration of the rigid body motion using the adjustable base was capable of generating distortions in the strain field. These distortions caused a lens that does not capture a completely flat image. This problem is most predominant at the edges of the field of view. Changing the objective lens for different magnifications does not seem to effect the distortion. By adding magnification at the rear of the lens assembly, much of the edge distortion can be reduced. However, this additional magnification reduces the maximum possible field of view the current system can achieve.

8.5 Conclusion

Early testing illustrated that DIC could detect local differences in material properties. Attempts to validate DIC measurements found that system noise was too high for values to be considered reliable. The project has explored different options for generating a speckle pattern at the desired scale, but none worked as well as the natural pattern on the polished aluminum. Southern Research has not fully explored the techniques for applying speckle with carbon powder and SEM stamp pads. It is possible better results could be generated with additional experimentation.

The project was able to measure the expansion of aluminum, but only using the natural surface not an applied speckle pattern. While these results were consistent with the known expansion of aluminum, there was significant noise in the system. The additional noise caused by the less than optimum speckle patterns made test results even less reliable.

The apparatus discussed has shown potential but is limited by the small field of view being attempted. Future work will need to identify larger material structures that need to be examined or material with higher thermal expansion. With a larger field of view, there will be more options for applying speckle to specimens and should be less sensitive to changes in focus. With a larger field of view, it may also be possible to mechanically strain the specimen rather than relying on thermal loading.

Acknowledgments Southern Research Institute. Cisco Systems, Inc.

References

1. Connolly, K.B., Ralph, W.C.: Elevated Temperature Optical Microscopy DIC, Advancement of Optical Methods & Digital Image Correlation in Experimental Mechanics, Volume 3. Conference Proceedings of the Society for Experimental Mechanics Series (2019)
2. Jones, I., Iadicola, M.: A Good Practices Guide for Digital Image Correlation. International Digital Image Correlation Society (2018)

Chapter 9
Perspective Compensation of 2D-DIC Measurements by Combination with Speckle Imaging

Juuso Heikkinen and Gary S. Schajer

Abstract Two-dimensional Digital Image Correlation (2D-DIC) is a popular method for determining object in-plane motion and deformation by tracing surface texture. However, the applicability of the method is limited by perspective errors that may be generated by object surface curvature, out-of-plane motions and non-perpendicular camera alignment. The resulting variations in camera magnification lead to errors in the measured surface motion signals.

Speckle imaging is a closely related DIC method where the motions of a laser-illuminated surface are tracked by analyzing the scattered interference speckle pattern. When the camera is in focus, the speckles move at the same rate as the surface. However, if the camera is defocused, the speckle motion rate deviates from the actual surface motion. Speckle imaging measurement sensitivity can thus be controlled by varying the amount of defocus. Interestingly, speckle imaging sensitivity increases with near-focus, in contrast to sensitivity reduction in 2D-DIC caused by perspective change. Therefore, if the surface motions are analyzed using both DIC and speckle imaging, the resulting sensitivity difference can be utilized to compensate any perspective errors present.

The amount of defocus must be moderated in order to achieve a balance between sensitivity and sufficient spatial resolution for full-field measurements. Therefore, it is important to study how speckle imaging behavior changes as a function of defocus. Here, the speckle imaging characteristics are experimentally investigated by recording speckle patterns at varying defocus distances and with different lens aperture sizes (f-numbers). The resulting speckle patterns are characterized by computing the average speckle diameters. The results reveal how at small defocus distances, speckle size is linearly dependent on lens f-number and inversely proportional to defocus distance. At large defocus distances, on the other hand, speckle size is independent on f-number and linearly proportional to defocus distance.

Keywords DIC · Perspective · Speckle imaging · Defocus · Autocorrelation

9.1 Introduction

Two-dimensional Digital Image Correlation (2D-DIC) is a simple and effective method for measuring object in-plane movements and deformations by tracking surface texture [1]. However, because 2D measurements contain no depth information, significant limitations are imposed on the measurable objects and motions. First, the specimen surface needs to be flat and aligned perpendicular to the camera. In addition, the object must remain at a fixed, uniform distance from the camera. These conditions enable the in-plane measurement sensitivity (magnification) to remain constant throughout the imaged area. However, in practice, measurement specimens often have non-flat surfaces and also move out-of-plane. The resulting variations in imaging distance cause corresponding magnification variations in the measured images, thus creating artifacts in the in-plane motion signals. Such perspective-induced effects could be minimized by using telecentric lenses, but such lenses are typically bulky and costly, especially for fields-of-view (FOV) larger than a few centimeters.

Speckle imaging is a variant DIC method where the laser speckles on the illuminated surface are used to track surface motions in place of painted speckles [2]. When the camera is focused at the surface, the imaged speckles move as if they were fixed to the surface. However, if the camera focal plane is shifted away from the specimen surface so that the camera becomes defocused, the movement of the observed speckles differs from the actual surface motion. At distances far away from the object, the observed speckle motions become larger with increasing defocus distance, contrary to the reduced magnification

J. Heikkinen · G. S. Schajer (✉)
Department of Mechanical Engineering, University of British Columbia, Vancouver, BC, Canada
e-mail: schajer@mech.ubc.ca

M.-T. Lin et al. (eds.), *Advancement of Optical Methods & Digital Image Correlation in Experimental Mechanics*,
Conference Proceedings of the Society for Experimental Mechanics Series, https://doi.org/10.1007/978-3-030-59773-3_9

in conventional imaging caused by perspective change. This gives the opportunity to use this characteristic to compensate for perspective effects. Furthermore, defocus in conventional imaging leads to blur that washes out surface texture, thus complicating the DIC analysis. Laser speckles, on the other hand, retain sharp contrast independent of defocus because the laser speckles have a 3D structure [3]. Therefore, the motivation for this project is to enhance conventional 2D-DIC analysis by using defocused speckle imaging to compensate for perspective changes and to maintain robust feature tracking.

9.2 Basics of Speckle Imaging

The speckle imaging method comprises two distinct types that have functional characteristics that can usefully be exploited in different specific applications. When a speckle pattern is captured using a focused camera, each point in the image corresponds to a specific point on the object (Fig. 9.1a). The associated speckle pattern, called "subjective," behaves as if it were attached to the surface. Subjective Speckle Imaging (SSI) is thus closely related to 2D-DIC and is useful for measuring surface deformation fields on objects that do not naturally possess strong surface texture.

In the case of the other variant, Objective Speckle Imaging (OSI), the speckle pattern is recorded by a bare, lensless, digital sensor. Because there is no lens that would focus light onto the sensor, each pixel receives light from all points within the illuminated surface, and any spatial information about the object surface is lost. OSI is thus applicable for measuring mainly rigid-body motions (RBM). However, unlike SSI, OSI can not only indicate surface displacements but can also measure surface rotations [2]. Furthermore, OSI measurement sensitivity increases with increasing recording distance, which makes it attractive for remote measurements.

Figure 9.1b shows the image formation principle of a defocused camera. When a conventionally imaged object is shifted further away from the focal plane, the magnification of the resulting image reduces. The image also becomes blurred, as

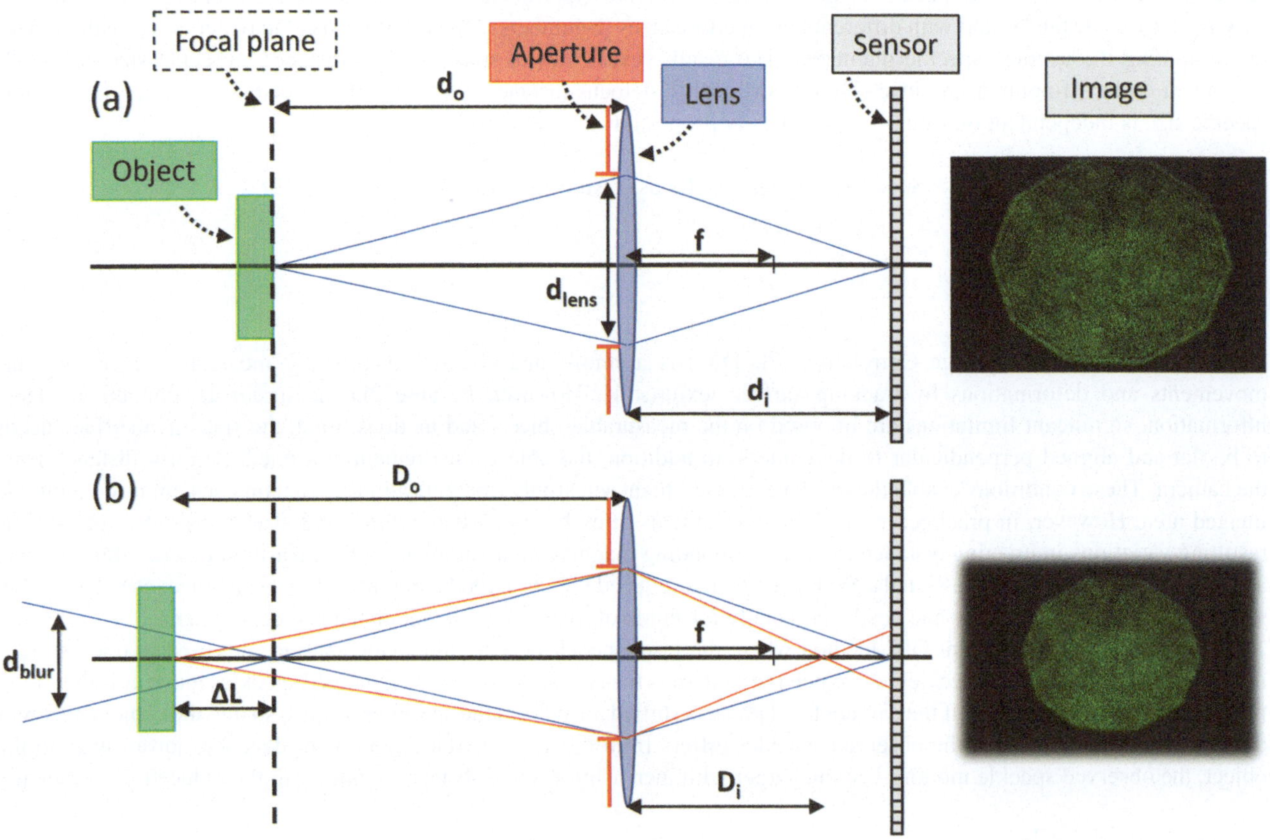

Fig. 9.1 Image formation in a simple optical system. (**a**) Object in-focus. (**b**) Defocused object. d_o, D_o = focused object distance, d_i, D_i = focused image distance, f = lens focal length, ΔL = defocus distance, d_{lens} = lens aperture diameter, and d_{blur} = blur circle diameter

indicated by the soft image edges. The blur occurs because a point on the sensor receives light from an extended surface area with a finite diameter d_{blur}.

Although not accurately focused on the specimen surface, a "defocused" camera is accurately focused at a focal plane in the adjacent space whose position depends on the camera focus adjustment. In the case of monochromatic laser illumination, a 3D speckle field exists in the space around the illuminated surface [4]. High speckle contrast is present at all spatial locations within that space, not just at the specimen surface. Consequently, a defocused camera accurately images a cross-section of the 3D speckle field that exists at the focal plane of the camera. In Fig. 9.1b, this corresponds to the line segment on the focal plane bounded by the red lines.

If the defocus distance ΔL is so high that the blur diameter reaches the size of the illuminated laser spot, then the resulting speckle pattern is exactly the same as would be observed by a bare (without lens) camera sensor placed in that plane. A highly defocused camera thus has the characteristics of OSI [3, 5]. A defocused camera has the advantage of not requiring to be placed physically within the speckle field. Instead, it may measure the speckle field remotely and "reach into" it according to the position of its focal plane. The sampling position can be varied simply by adjusting the camera (de)focus [5].

9.3 Research Objective

While OSI is attractive for measuring rigid body motions at high sensitivity, the inherent loss of spatial resolution makes it impractical for full-field measurement applications where 2D-DIC is commonly used. Therefore, the interest in this research is to study the intermediate region between Subjective and Objective Speckle Imaging. This region mixes characteristics of both regimes and allows them to be used simultaneously.

If speckle imaging is performed using moderately defocused camera, then each point on the sensor receives light from limited, small area on the object surface, and the resulting speckle patterns maintain reasonable spatial resolution. In Fig. 9.1b, the intermediate range is characterized by small defocus distances so that the extent of blur on the object is much smaller than object dimensions. However, defocused speckle imaging differs from 2D-DIC in that its sensitivity increases with defocus distance rather than decrease. Therefore, if the in-plane motions of a near-focused object are measured using both DIC and speckle imaging, the resulting sensitivity difference can be used to compensate for the effects of perspective changes.

9.4 Speckle Pattern Characterization

There is a trade-off between spatial resolution and sensitivity, so for effective use, the character and extent of the intermediate region between Subjective and Objective Speckle Imaging needs to be well understood. The average speckle diameter is a statistical parameter that characterizes the speckle imaging geometry. Importantly, the subjective speckle size is governed by different geometric parameters than objective speckle size. Therefore, determining the speckle size as a function of defocus distance provides a practical way to assess the transition from subjective to objective speckles, as well as the extent of the intermediate regime of moderate defocus.

The diffraction limited subjective speckle diameter can be defined as [6]

$$S_{\text{subj}} = 1.22\,\lambda\,(1+M)\,\frac{f}{d_{\text{lens}}} = 1.22\,\lambda\,(1+M)\,f\# \qquad (9.1)$$

where λ is the laser wavelength, f is the camera lens focal length, d_{lens} is the lens aperture diameter, $f\# = f/d_{\text{lens}}$ is the lens f-number, $M = d_i/d_o$ is the camera magnification, d_i is the image distance, and d_o is the object distance.

The diffraction-limited objective speckle diameter is [6]

$$S_{\text{obj}} = 1.22\,\lambda\,M\,\frac{\Delta L}{d_{\text{spot}}} \qquad (9.2)$$

where ΔL is the defocus distance and d_{spot} is the diameter of the illuminated spot. For objective speckle imaging setup, ΔL is the object–sensor distance and $M = 1$. If a highly defocused camera is used instead, ΔL is the object–focal plane distance, and M is the camera in-focus magnification ratio.

According to Eq. (9.1), the subjective speckle size depends linearly on the lens numerical aperture. Therefore, the ratio $S_{subj}/f\#$ should be a constant when $\Delta L = 0$. Conversely, at large defocus distances, the observed speckle size should be independent of the lens aperture and scale linearly with ΔL.

Based on the geometry shown in Fig. 9.1, the blur diameter depends on both the defocus distance and the aperture diameter according to relation:

$$\frac{d_{lens}}{d_o} = \frac{d_{blur}}{\Delta L} \tag{9.3}$$

Therefore, the minimum defocus distance required for the speckle size to obey the theoretical Equation (9.2) must be such that the blur diameter equals the illuminated spot diameter, $d_{blur} = d_{spot}$. Thus, the minimum defocus distance is

$$\Delta L_{min,\ obj} = \frac{d_o}{d_{lens}} d_{spot} = \frac{d_o d_{spot}}{f} f\# \tag{9.4}$$

9.5 Speckle Size vs. Aperture Size vs. Defocus

The speckle size dependence on aperture size and defocus distance was experimentally investigated. The object was a flat medium-density fiberboard (MDF) plate with a diffuse surface. A single-mode green laser source (JDS Uniphase, $\lambda = 532$ nm) with a diverging beam illuminated the object surface at a 31° angle with respect to the surface normal. The illuminated spot size was controlled by placing an aperture element in front of the laser so that only the central portion of the beam ($d_{spot} = 19.5$ mm) with approximately uniform intensity profile passed through.

The scattered interference speckle pattern was imaged using a Canon EOS 100D DSLR camera (resolution 5184x3456 pxl^2, pixel diameter $d_{pixel} = 4.3$ μm) fitted with a Canon EF $f = 50$ mm $f\#1.8$ lens. A 34 mm extension tube was attached between the camera body and the lens to reduce the focus distance, giving a magnification $M = 0.83$. The interference speckle pattern was sampled at different defocus distances ranging from –20 mm (far focus, focal plane behind the object surface) to 1350 mm (near focus, focal plane in front of the object surface). For each defocus distance, the speckle pattern was imaged using three different numerical apertures, $f\#5.6$, $f\#11.0$, and $f\#22.0$. The defocus distance was changed by moving the camera along an optical rail rather than changing the lens focus settings. This arrangement maintained constant magnification throughout the measurements. Figure 9.2 shows the experimental setup and an example speckle pattern.

For each speckle image, the average speckle size was determined using a two-dimensional normalized autocorrelation, implemented in Matlab. The autocorrelation window consisted of a cropped 201 × 201 pxl^2 region extracted from the image

Fig. 9.2 (Left) Experimental setup. (Right) Example speckle pattern captured at $\Delta L = 0$ and $f\#22.0$

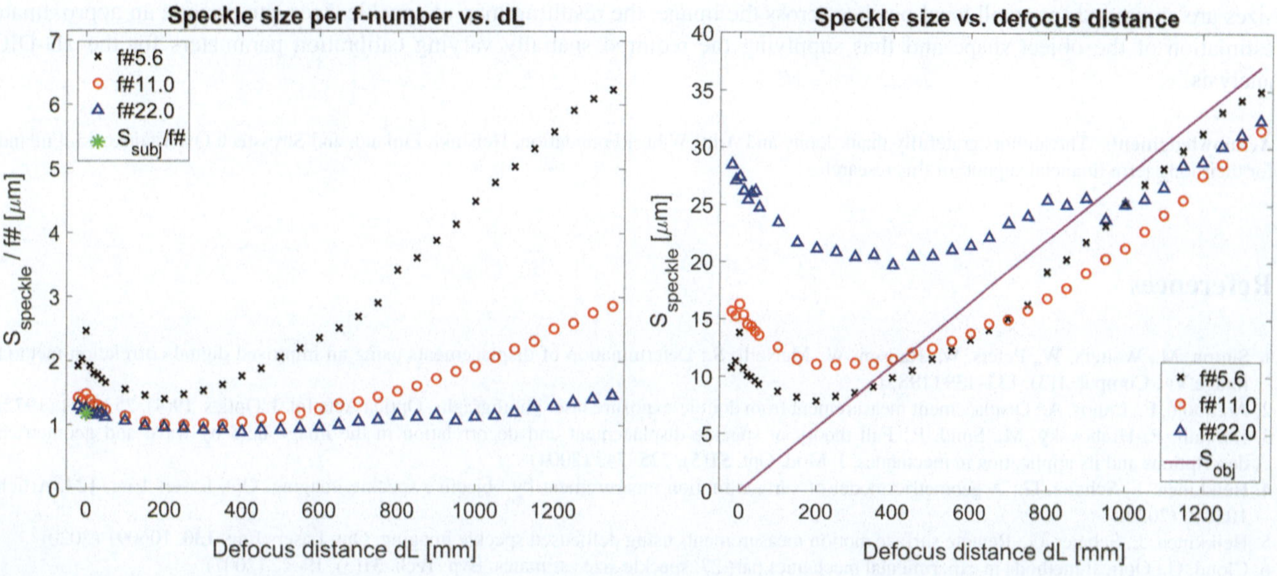

Fig. 9.3 (Left) Measured speckle size per f-number vs. defocus distance. (Right) Measured speckle size vs. defocus distance

center. The resulting autocorrelation map contains a central peak whose width at half-height represents the statistical average speckle size. Figure 9.3 (Left) shows the measured speckle sizes per f-number for different defocus distances, as well as the theoretical subjective speckle diameter per f-number calculated using Equation (9.1). The speckle patterns recorded using *f*#11.0 and *f*#22.0 have values that are very close to the theoretical expectation, while the speckle size recorded at *f*#5.6 was higher than expected. Figure 9.3 (Right) shows a comparison between the measured speckle sizes and the theoretical objective speckle size determined from Equation (9.2) using the specific illumination and imaging parameters. At large defocus distances, the measured speckle sizes converge close to the theoretical line, independent of the aperture size.

9.6 Discussion

It is possible to distinguish three different regions in the speckle size graphs. At small defocus distances (0-100 mm), perspective effects cause the observed speckle size to be inversely proportional to defocus distance. When the camera moves further away, the effective magnification reduces, causing the imaging scale to become smaller, similar to 2D-DIC. At very large defocus distances (hundreds of mm's), on the other hand, the speckle size increases linearly with defocus distance, similar to objective speckle imaging. Between these two extremes, at moderate defocus distances (around 200-400 mm), there is a transition region where the speckle size reaches minimum values. This can be explained by noting that each point on the sensor plane of a defocused camera receives light from a finite-area blur circle whose diameter defines the effective spot diameter. Because the blur circle increases linearly with defocus distance, the effective spot size expands until the blur diameter reaches the physical spot diameter. In this regime, the ratio $\Delta L/d_{\text{spot}}$ in Equation (9.2) remains almost constant, leading to the flat section around the minimum speckle diameter position. Because blur diameter is inversely proportional to the f-number, the effective spot diameter of *f*#5.6 reaches the physical spot diameter at smaller defocus distances than *f*#11.0 or *f*#22.0, as indicated by Equation (9.4). Correspondingly, *f*#5.6 begins to correlate with the theoretical objective speckle diameter at smaller defocus distances than the two smaller apertures.

9.7 Conclusion

The linear dependence between the speckle size and defocus at small defocus distances is valuable for perspective compensation because a reduction in speckle size can be related to an increase in object distance. In addition, this regime is the most practical range for full-field measurements, since moderate defocus maintains high spatial resolution. If the speckle sizes are computed for small local patches across the image, the resulting map of speckle sizes can provide an approximate estimation of the object shape and thus supplying the required spatially varying calibration parameters for the 2D-DIC analysis.

Acknowledgments The authors gratefully thank Jenny and Antti Wihuri Foundation, Helsinki, Finland, and Stresstech Oy, Vaajakoski, Finland, for their long-term financial support of this research.

References

1. Sutton, M., Wolters, W., Peters, W., Ranson, W., McNeill, S.: Determination of displacements using an improved digital correlation method. Image Vis. Comput. **1**(3), 133–139 (1983)
2. Archbold, E., Ennos, A.: Displacement measurement from double-exposure laser photographs. Optica Acta Int. J. Optics. **19**(4), 253–271 (1972)
3. Horvath, P., Hrabovský, M., Smid, P.: Full theory of speckle displacement and decorrelation in the image field by wave and geometrical descriptions and its application in mechanics. J. Mod. Opt. **51**(5), 725–742 (2004)
4. Heikkinen, J., Schajer, G.: A geometric model of surface motion measurement by objective speckle imaging. Opt. Lasers Eng. **124**, Article 105850 (2020)
5. Heikkinen, J., Schajer, G.: Remote surface motion measurements using defocused speckle imaging. Opt. Lasers Eng. **130**, 106091 (2020)
6. Cloud, G.: Optical methods in experimental mechanics part 27: speckle size estimates. Exp. Tech. **31**(3), 19–22 (2007)

Chapter 10
Holographic Measurement of Semi-transparent Tympanic Membrane Shape Using Multiple Angle Illuminations

H. Tang, P. Psota, J. J. Rosowski, J. T. Cheng, and C. Furlong

Abstract The shape of the tympanic membrane (TM) plays an important role in conducting acoustic energy from the environment to the ear for hearing. Changes of the TM shape also have the diagnostic value, e.g., bulging of the TM is a sign of middle ear effusion. Previously we developed a high-speed holographic system employing a tunable wavelength laser for rapid TM shape measurement. However, because of the semi-transparency of the TM and the short-exposure time required for the high-speed acquisition, the illumination intensity (0.01 mW/mm^2) of the tunable laser was insufficient for measurements on the semi-transparent TM surface. Here, we describe a multi-angle illumination technique that allows us to use a single wavelength (532 nm) laser with higher illumination intensity (0.06 mW/mm^2) to perform shape measurements of the semi-transparent TM. The accuracy of the proposed method is demonstrated by measurements of a stepwise gauge provided by the National Institute of Standards and Technology. We successfully applied the abovementioned method to resolve the shape of the fresh postmortem human TM. We see a potential for miniaturization of the apparatus into a holographic otoscope for in-vivo measurements.

Keywords High-speed holography · Human tympanic membrane · Single wavelength shape measurement

10.1 Introduction

Full field of view and noninvasive measurement of the surface shape of the tympanic membrane (TM) with high resolution is of interest for hearing research. The vibration of the TM initializes sound energy transmission from the environment to the inner ear [1, 2]. The TM's complex geometry (shape and thickness) affects the energy transmission [3] and is essential to understanding the mechanics of the TM. The shape information also has diagnostic values for middle ear pathologies, e.g., bulging of the TM is a sign of middle ear effusion. Our group has developed a high-speed multiple wavelength Digital Holographic (DH) system using a tunable laser to measure the TM shape in cadaveric human ears [4, 5]. However, the tunable laser illumination is not sufficient to measure the shape of the TM without surface preparation due to the semi-transparency of the TM and short exposure time for high-speed image acquisition. To overcome this limitation and enable the high-speed holographic system for in vivo measurement, we apply a high-power single wavelength laser to resolve the shape with a multiple angle DH method [6] together with a temporal phase unwrapping approach to perform robust and

H. Tang (✉)
Center for Holographic Studies and Laser micro-mechaTronics (CHSLT), Worcester Polytechnic Institute, Worcester, MA, USA

Mechanical Engineering Department, Worcester Polytechnic Institute, Worcester, MA, USA
e-mail: htang3@wpi.edu

P. Psota
Faculty of Mechatronics, Informatics and Interdisciplinary Studies, Technical University of Liberec, Liberec, Czech Republic

J. J. Rosowski · J. T. Cheng
Eaton-Peabody Laboratory, Massachusetts Eye and Ear Infirmary, Boston, MA, USA

Department of Otology and Laryngology, Harvard Medical School, Boston, MA, USA
e-mail: John_Rosowski@meei.harvard.edu; Tao_Cheng@meei.harvard.edu

C. Furlong
Mechanical Engineering Department, Worcester Polytechnic Institute, Worcester, MA, USA
e-mail: cfurlong@wpi.edu

M.-T. Lin et al. (eds.), *Advancement of Optical Methods & Digital Image Correlation in Experimental Mechanics*,
Conference Proceedings of the Society for Experimental Mechanics Series, https://doi.org/10.1007/978-3-030-59773-3_10

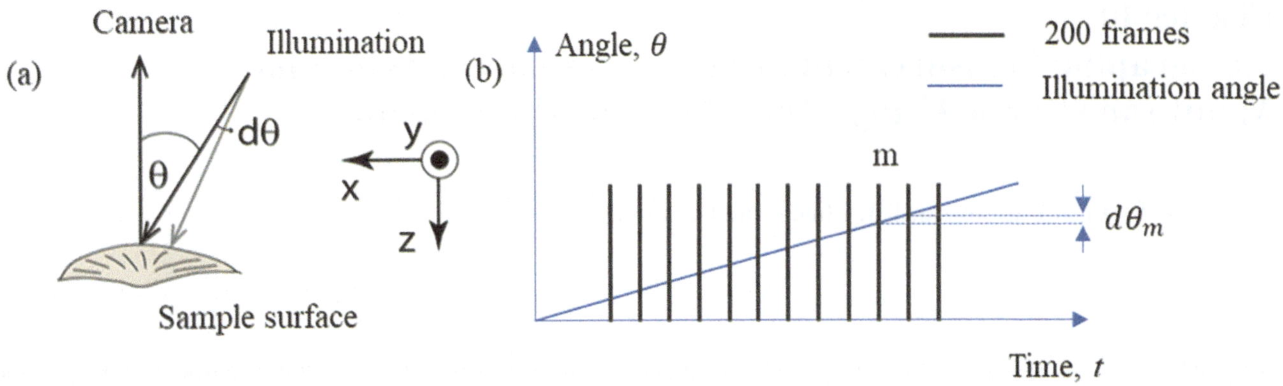

Fig. 10.1 (**a**) Holographic setup for shape measurement. (**b**) Timeline of the event for shape measurement

accurate shape measurement on the semi-transparent TM surface. Accuracy of the new method is estimated by a measure of a stepwise gauge provided by the National Institute of Standards and Technology (NIST). We successfully applied the new shape measurement method on a fresh postmortem human TM without any surface preparation.

10.2 Methods

Digital holography applies optical sensors (cameras) to record the optical path difference of the reference beam and object beam from a coherent light source. It can reconstruct the optical wavefield (amplitude and phase) scattered by the measured object. If a controlled change of illumination angle $d\theta$ is introduced to the holographic setup shown in Fig. 10.1a, the contour of the sample's shape (in the z-direction) has a relationship with the wavelength λ and the illumination angle θ as follows [7]:

$$dP = \frac{2\pi}{\lambda} \sin\theta z \, d\theta, \tag{10.1}$$

where dP is the optical phase difference due to the change of the illumination angle.

To increase the robustness of the measurement and avoid spatial phase unwrapping errors, we introduced a continuous change of the illumination angle and capture serials of high-speed images at discrete timings during the change. The timelines of changing the illumination and camera capture are shown in Fig. 10.1b. Each black line represents one serial of 200 frames of high-speed image acquisition at 67200 frame per second, and the blue line is the illumination angle. Because each serial of high-speed imaging is much faster (0.03 s) than the motion of the illumination change (1 s), the θ is assumed to be constant during the individual serial of high-speed image acquisition. Temporal phase unwrapping is performed by calculating the cumulative sum of the consecutive optical phase difference to solve the total optical phase dP_m difference between the first and last serials of high-speed image acquisitions [7]:

$$dP_m = \sum_{m=1}^{m} \frac{2\pi}{\lambda} \sin\theta z \, d\theta_m \tag{10.2}$$

where dP_m is the total optical phase difference due to the m step changes of the illumination angle.

10.3 Results

A NIST traceable gauge is used to determine the measurement resolution. An image of the gauge is shown in Fig. 10.2, and the measured area is shown in the zoomed view. The minimum height change of the target is 0.25 mm (1.5–1.25 mm).

A 3D mesh plot of the NIST gauge and the measured shape of the gauge along the red line are shown in Fig. 10.3a and b. The shape result suggests the shape measurement can distinguish height difference as small as 0.25 mm and surface jump as large as 1.5 mm. The histogram of the red line in Fig. 10.3b and the standard deviation at each height are shown in Fig. 10.4. The measured results are normally distributed around three significant peaks corresponding to individual steps of the NIST gauge with a maximum standard deviation of 0.044 mm.

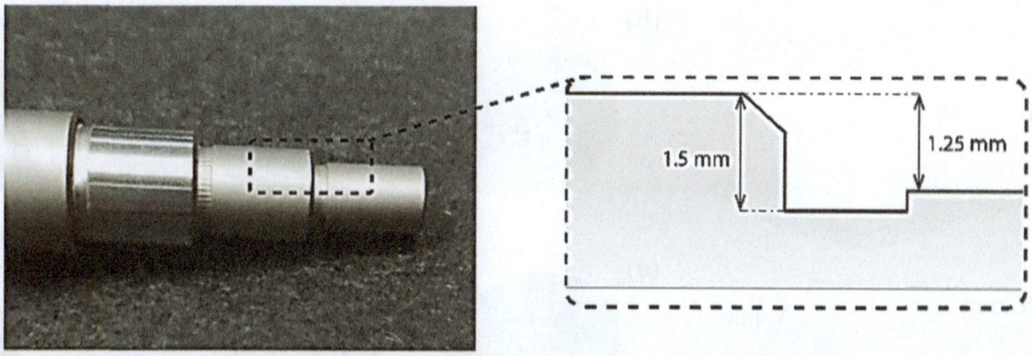

Fig. 10.2 Dimension of the NIST gauge. The measured area is shown in the zoomed view within the dashed line

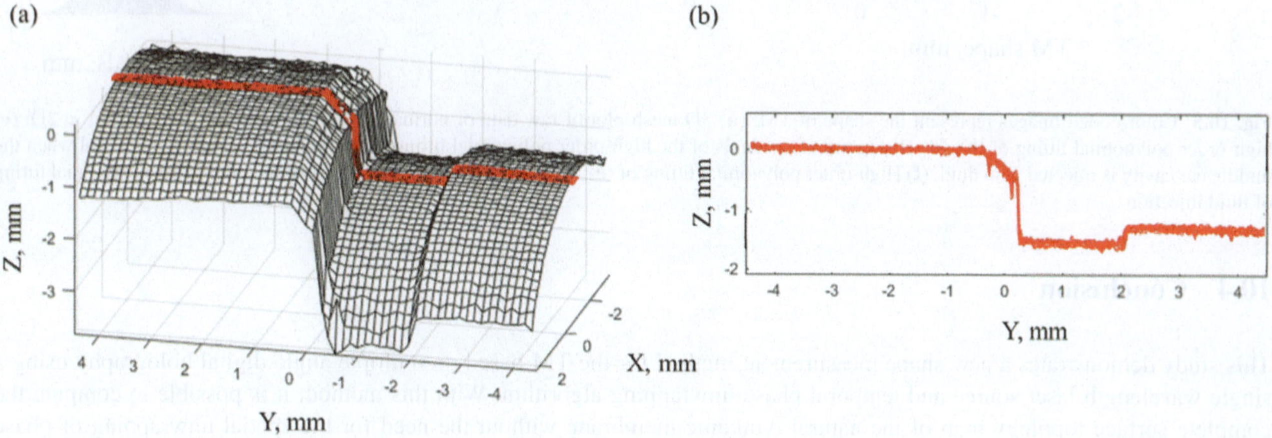

Fig. 10.3 (**a**) A 3D mesh plot of measured NIST gauge. (**b**) 1D data of the redline in (**a**) along the X-axis

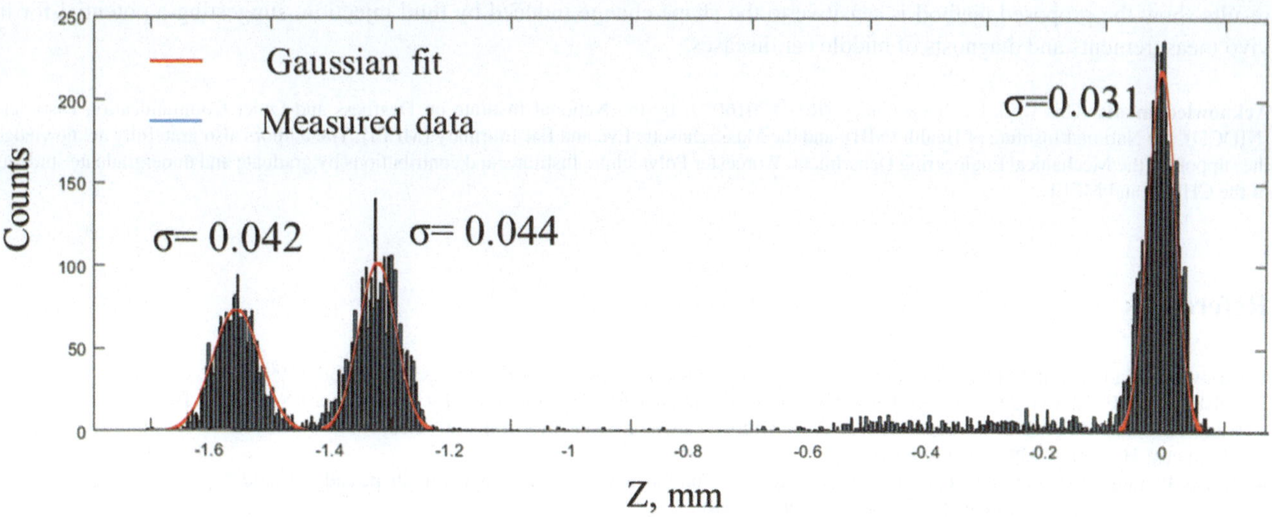

Fig. 10.4 Histogram of the measured data marked red in Fig. 10.3b and the Gaussian fits at each peak

The shape of a fresh cadaveric TM without any surface painting is measured using the multi-angle shape measurement system before and after injecting fluid into the middle ear cavity. The shape measurement result before the fluid injection is shown in the 3D mesh plot in Fig. 10.5a. Figure 10.5b and c is the 2D presentation of the raw data and the related high-order polynomial surface fitting before injecting fluid. The residual of the two is shown in a histogram in Fig. 10.5d with a standard deviation of 0.07 mm. Figure 10.5e and f shows the raw data and surface fitting after injecting fluid into the middle ear cavity with a standard deviation of 0.125 mm shown in (g). Injecting fluid into the middle ear cavity results in a bulging TM in the area (yellow coded) where the surface of the TM is in contact with the fluid.

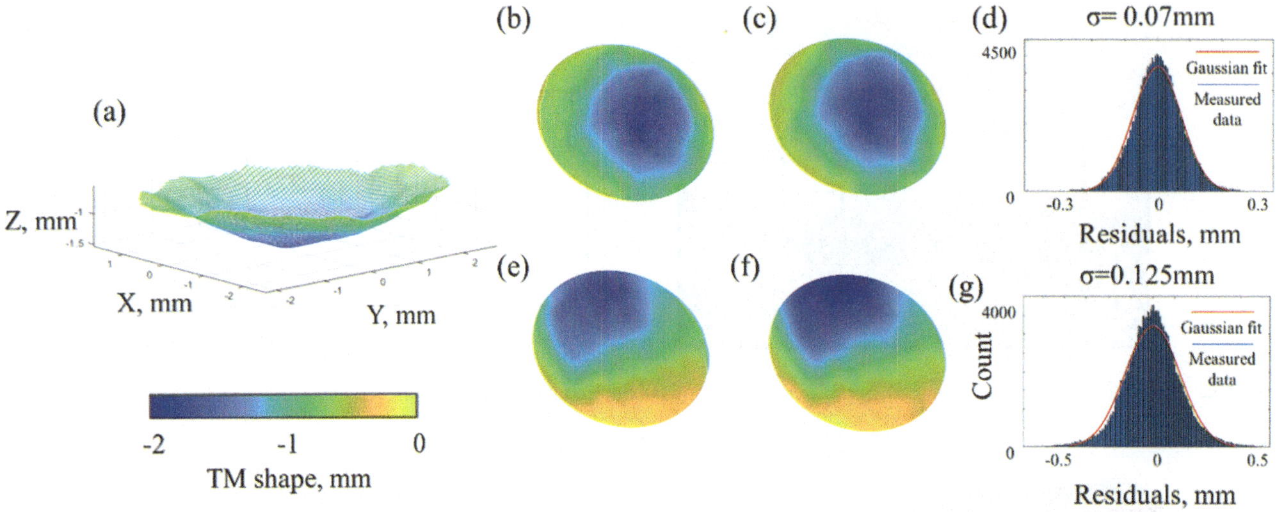

Fig. 10.5 Color-coded images represent the shape of TM: (**a**) 3D mesh plot of raw data of normal TM; (**b**) raw data of normal TM in 2D; (**c**) high-order polynomial fitting of (**b**). (**d**) Histogram of residuals of the high-order polynomial fitting of normal TM; (**e**) raw data of TM when the middle ear cavity is injected with fluid. (**f**) High-order polynomial fitting of (**e**). (**g**) Histogram plot of residuals of the high-order polynomial fitting of fluid injection

10.4 Conclusion

This study demonstrates a new shape measurement method for the TM based on multiple angle digital holography using a single wavelength laser source and temporal phase unwrapping algorithm. With this method, it is possible to compute the complete surface topology map of the natural tympanic membrane without the need for the spatial unwrapping of phase difference maps. The method was validated by measuring a stepwise gauge provided by NIST with known dimensions. A fresh non-fixed human postmortem tympanic membrane was measured with and without middle ear fluid injection. The results show the proposed method is sensitive to the shape change induced by fluid injection, suggesting a potential for in vivo measurements and diagnosis of middle ear diseases.

Acknowledgments This work has been funded (R01DC016079) by the National Institute on Deafness and Other Communication Disorders (NIDCD), the National Institute of Health (NIH), and the Massachusetts Eye and Ear Infirmary (MEEI). The authors also gratefully acknowledge the support of the Mechanical Engineering Department, Worcester Polytechnic Institute, and contributions by graduate and undergraduate students of the CHSLT and MEEI.

References

1. Rosowski, J.J.: Outer and Middle Ears Comparative Hearing: Mammals, pp. 172–247. Springer, New York (1994)
2. Geisler, C.D.: From Sound to Synapse: Physiology of the Mammalian Ear. Oxford University Press, Oxford/New York (1998)
3. Wang, X., Guan, X., Pineda, M., Gan, R.Z.: Motion of tympanic membrane in guinea pig otitis media model measured by scanning laser Doppler vibrometry. Hear. Res. **339**, 184–194 (2016)
4. Razavi, P., Tang, H., Rosowski, J.J., Furlong, C., Cheng, J.T.: Combined high-speed holographic shape and full-field displacement measurements of tympanic membrane. J. Biomed. Opt. **24**(3), 031008 (2018)
5. Tang, H., Razavi, P., Pooladvand, K., Psota, P., Maftoon, N., Rosowski, J.J., Furlong, C., Cheng, J.T.: High-speed holographic shape and full-field displacement measurements of the tympanic membrane in normal and experimentally simulated pathological ears. Appl. Sci. **9**(14), 2809 (2019)
6. Ichirou, Y., Ohta, S., Kato, J.: Surface contouring by phase-shifting digital holography. Opt. Lasers Eng. **36**(5), 417–428 (2001)
7. Psota, P., Tang, H., Pooladvand, K., Vít, L., Furlong, C., Rosowski, J.J., Cheng, J.T.: Investigation of tympanic membrane shape using digital holography. Optics Measur. Int. Conf. **11385**, 113850G (2019)

Chapter 11
Characterization of Interface Debonding Behavior Utilizing an Embedded Digital Image Correlation Scheme

Tomislav Kosta and Jesus O. Mares Jr.

Abstract We present an in situ Digital Image Correlation (DIC) technique to characterize the interface debond behavior in a Sylgard 184-based particulate composite under quasi-static tensile loading. We construct tensile specimens with a single 650 μm glass bead inclusion at the center of the sample along with an embedded DIC speckle pattern at the midplane to enable measurement of the strain field in the region surrounding the embedded glass bead. The planar nature of the speckle pattern paired with the transparent binder allows for measurement of the sub-surface strain with simple camera diagnostics. Symmetry of the tensile specimen enables straightforward interpretation of the results. By capturing the evolution of the strain field from the unloaded state up to, and through, the delamination, we obtain a rich data set describing the debond process. Through repetition of the experiment, the debond behavior can be described in a statistically meaningful manner. This approach allows for calibration of a targeted constitutive model and serves as a tool to probe the effect of material or surface modification to promote adhesion.

Keywords Sylgard 184 · Interfaces · Interface debonding · Interface delamination · DIC

11.1 Introduction

The mechanical properties of composite materials depend on both the properties of the individual components as well as the properties of the interfaces between those components, particularly in the context of material strength. Depending on the filler volume fraction, the interface debond strength can have a significant influence on the strength of the composite. We refer to the interface debond as the mechanical separation of two separate components in a composite material, or delamination. The interface debonding strength (or adhesion strength) becomes particularly important for the development of high fidelity computational models to simulate events in which strength and flow properties play a role. However, accurate statistical representation of the delamination occurring within composite materials under mechanical loading is difficult to achieve due to the technical challenges in the characterization of the interface properties.

Several standardized test methods exist for the measurement of adhesion of a simple binary material system; however these tests are poorly suited for quantitative assessment of the interface debonding strength in composite materials. Tests such as the pull [1], peel [2], and blister test [3] provide a means of quantifying the debonding process; however, the stress states of the materials in these experiments are non-ideal and complicate the interpretation of the effective properties associated with delamination. These tests tend to produce data which is highly dependent upon the specific geometry of the debonding process, proving difficult to characterize the general interface properties. Additionally, due to required sample geometries, these tests are not well-suited to examine particulate bulk composites, making them difficult to apply to many material systems of interest.

Several methods have been developed to investigate the effective interface debonding strength by examining the fracture of particulate bulk composites. For example, Tan et al. [4] have integrated digital image correlation (DIC) into a mode-I fracture test of a composite sample to provide targeted data for cohesive zone element model parameterization. These experiments provide direct results for model calibration; however, the delamination properties that dominate the delamination occurring at individual particle-binder interfaces are heavily integrated through this approach. The resultant models are

T. Kosta (✉) · J. O. Mares Jr.
United States Air Force Research Laboratory, Munitions Directorate, Eglin AFB, FL, USA
e-mail: tomislav.kosta@us.af.mil; jesus.mares.2@us.af.mil

M.-T. Lin et al. (eds.), *Advancement of Optical Methods & Digital Image Correlation in Experimental Mechanics*, Conference Proceedings of the Society for Experimental Mechanics Series, https://doi.org/10.1007/978-3-030-59773-3_11

typically calibrated to define single valued parameters which lose the capability to capture the statistical distribution of the interface properties throughout the bulk material.

Alternatively, experiments such as those by Gent and Park [5] and proposed by Lauke [6, 7] explicitly study the delamination of a single particle within a binder matrix. Specifically, Gent and Park [5] examined the delamination of a glass sphere within an optically transparent binder material under an applied tensile load and observed the onset of debonding at the interface as well as additional features such as the cavitation or crazing of the binder near the particle-binder interface. The stress states within the binder material which generated the debond event were analytically determined and noted as the critical debonding stress. This critical debonding stress was noted to serve as an indicator of the strength of adhesion between the particle and the binder material.

In this work we extend the methodology introduced by Gent and Park [5] to incorporate the Digital Image Correlation (DIC) technique to quantify the strain fields of the binder material near an embedded glass bead through a debonding event. This method can serve as a basis to collect a statistically relevant set of debonding information for a coupled binder and particle system of interest. The data acquired with this technique allows for enhanced validation of modeling approached to ensure proper capture of the salient physics.

11.2 Experimental Methods

11.2.1 Sample Fabrication

To allow for the inference of the strain state of the binder material during a debonding event, standard tensile test samples were constructed of a transparent matrix material with a single spherical particle inclusion surrounded by a plane of fine tracking particles. Rectangular dog-bone tensile samples with a cross section approximately 9 mm by 12 mm were constructed from Sylgard 184 in a layered approach in order to properly embed a spherical glass particle which serves as the inclusion, as well as a plane of tracking particles for use in DIC. Dow Corning Sylgard 184 silicone elastomer was used for the binder material and was mixed in a 10:1 base to curative ratio by mass for all samples. The highly transparent nature of the material allowed for the observation of particles and debonding within the sample throughout the experiment. An initial layer of Sylgard 184 was poured into a dog-bone shaped mold to a height of approximately 6.35 mm and cured at 100 °C for 30 min. A thin layer of binder was then applied to the partially cured surface, and a spherical glass particle was placed in the middle of the sample such that the thin layer of binder was level with the midplane of the particle. In order to allow for suitable observation of the delamination event, glass particles with a nominal diameter of 0.65 mm were selected from a lot of 3 M hollow glass microspheres. The sample was then cured for an additional 30 min at 100 °C. A thin layer of a mixture of Cospheric 27–32 μm black paramagnetic polyethylene microspheres and Sylgard 184 was then applied to the surface of the new cure layer to serve as the DIC speckle pattern, and the mold was then filled to the full mold height of 12.70 mm. The sample was then cured at 100 °C for 30 min. The sample was released from the mold and excess binder was trimmed. The sample construction process is summarized in Fig. 11.1, and a notional diagram of the fully constructed sample is presented in Fig. 11.2.

The quality of the placement of the spherical glass particle and the surrounding plane of tracking particles were evaluated with the use of a Zeiss Discovery V12 stereoscope. The glass particles were examined to ensure their placement at the midplane through the thickness of the dog-bone test article, as well as at the mid-location of the sample in the front view. As the success of the strain field measurement using DIC is highly dependent upon the quality of the application of the

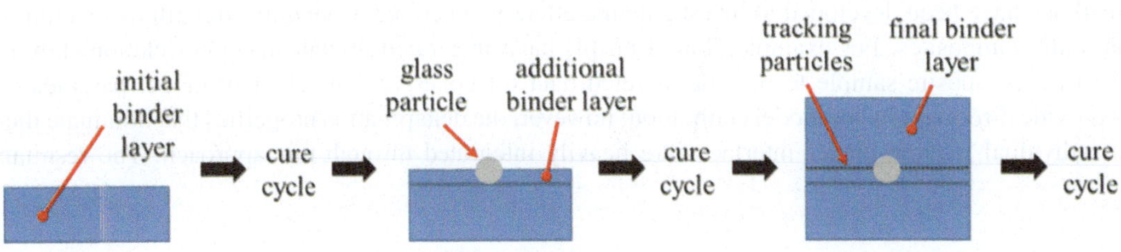

Fig. 11.1 Illustration of the layered approach to construct dog-bone samples of Sylgard 184 with an embedded spherical glass particle and pattern of tracking particles

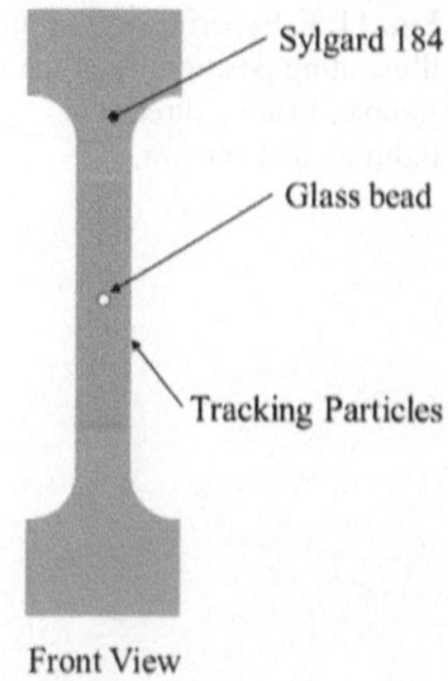

Fig. 11.2 Front view of tensile load test sample with embedded glass bead and digital image correlation speckle pattern

tracking pattern [8], a number of different tracking particles were used in a series of initial samples in order to down-select to a suitable tracking pattern. It was found that Cospheric 27–32 μm black paramagnetic polyethylene microspheres served as ideal tracking particles for this experiment setup.

11.2.2 Mechanical Testing

A global tensile stress condition was imparted upon the sample through the use of a conventional load frame, as conducted in a typical tensile test. An MTS Exceed E42.503 (5 kN) load frame with a 1kN load cell was used to apply a constant velocity extension while also measuring a continuous tensile load to the sample. The load frame was controlled by MTS TestSuite Elite software. The extension of the load frame cross head was held at a constant rate of 0.5 mm/s in order to apply a constant global engineering strain rate to the sample. The sample that was preloaded at a value of 5 N of tension to properly orient the sample is the grips and position the camera. The glass particle and surrounding tracking pattern was imaged with the use of a Point Grey Grasshopper 2 camera with a resolution of 2048 by 2048 pixels. A Rokinon 2.8/14 mm optical lens was used in concert with the camera to capture highly detailed images of the region of interest which corresponds to a field of view of approximately 10 by 10 mm with an approximate standoff distance of 0.5 m. Illumination of the sample was supplied by the use a Visual Instrumentation Corp. 900420 LED module lamp. Given the transparent nature of Sylgard 184, lighting may be accomplished by direct illumination or by backlighting. After attempting both methods, direct lighting, where the illumination source was on the same side of the sample as the camera, was found to allow for greater success with the DIC analysis.

Due to the load frame configuration, the load is applied to the sample with the controlled displacement rate of an upper grip, while the lower grip is held stationary. This configuration subsequently results in a displacement of the glass particle nominally equal to one half of the displacement of the crosshead controller of the load frame. This behavior in turn makes it difficult to continuously observe the glass particle and surrounding tracking pattern at high resolution. In order to counter this effect, the camera was mounted to a Thorlabs LTS150 motorized stage in the vertical direction and controlled with Thorlabs Kinesis software. The displacement of the stage was then programmed to travel at a constant rate equal to one half that of the travel rate of the crosshead controller. An image of the experimental setup is shown in Fig. 11.3.

Images were captured using Point Grey FlyCapture 2 software, and the loading conditions were measured such that the imaging and loading information of the sample were recorded as a function of common time. This methodology was successful in capturing the relative translation of the tracking particles with respect to the embedded glass particle and clearly captured the delamination process between the binder and particle. The captured images of a sample before and after delamination are shown in Fig. 11.4.

11.3 Data Analysis and Results

To determine the strain state of the binder material near the glass inclusion throughout the experiment, the acquired images were processed using Correlated Solutions Vic-2D Digital Image Correlation software, version 7. A parameterization study

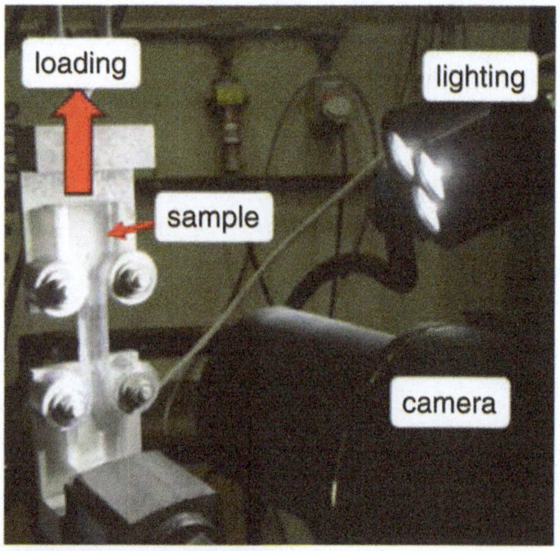

Fig. 11.3 Experimental setup illustrating positioning of the sample, loading direction, lighting, and camera

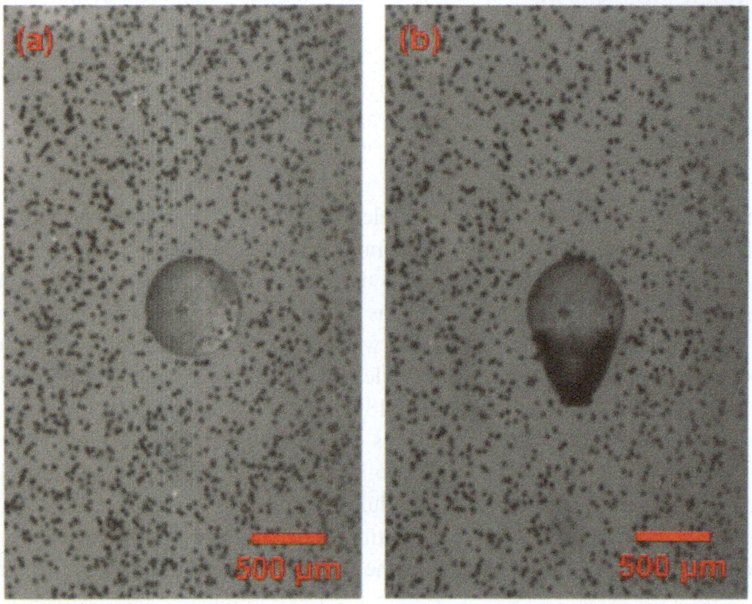

Fig. 11.4 Image captured during tensile loading of sample (**a**) before and (**b**) after delamination event

was conducted to identify a set of appropriate parameter specifications. The parameters of primary interest included step size, noise level, and the interpolation settings. It was determined that the following set of parameters allowed for the greatest performance of the calculation of the strain field: a subset value of 39, noise level of 2.00, step size of 4, Gaussian subset weights, an optimized 8-tap interpolation, a criterion of squared differences, and no low-pass filter, with an incremental correlation and exhaustive search.

The DIC software was successful at determining the spatially resolved strain maps as a function of time up to the point of, and shortly after, delamination between the particle and glass inclusion. A series of processed strain results obtained through this analysis is shown in Fig. 11.5.

Due to the known capture rate of the camera and the data acquisition rate of the load frame, a common time was established to connect the DIC information to the global load conditions imparted upon the sample. Therefore, the global tensile load and extension of the sample is known at the point of debonding. Additionally, this information coupled with the calculated local strain information near the embedded glass particle and the spatial information of the resulting void structure provide a rich data set of the debonding process. This information can then be used to calibrate and validate modeling and simulation efforts to capture the process of a debonding event. Additionally, the stress state of the binder material can be determined through the known stress-strain relationship of the material.

Fig. 11.5 A series of images and calculated ε_1 strain through the tensile loading test up to and past the point of a forced delamination event achieved via DIC

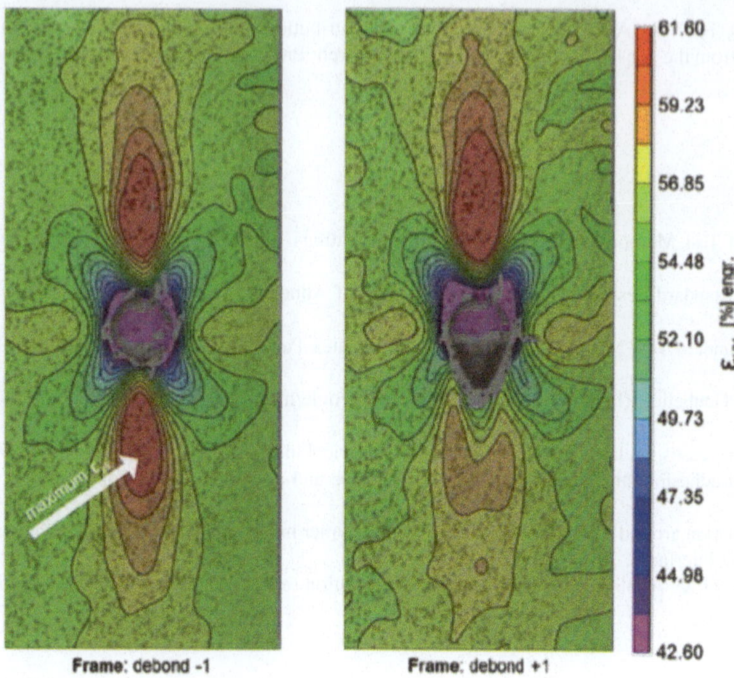

Fig. 11.6 Calculated strain field near the embedded particle in the –yy direction one frame before and one frame after the debonding event. The location of the maximum local tensile strain is highlighted

However, using the available strain information from the DIC analysis results, the maximum local tensile strain in the direction of the applied load was used as a metric to quantify adhesion between the particle and binder material. Due to the geometry of the loading condition, the observed debonding processes is most heavily dominated by the forces in the pure tensile direction at the interface, which corresponds to the strain in the –yy direction obtained from the analysis. It was noted that the maximum local strain value was taken on the side of the particle which leads to the initial debond, as debonding can occur on either side (top or bottom) of the particle in the direction of the applied load. This location of maximum strain taken at the frame before the observed debonding event is highlighted in Fig. 11.6.

A collection of results for a significant number of debonding events can be aggregated for a specific particle and binder material to represent an effective distribution of adhesion properties which govern the debonding process between the two materials. This technique could thereby be used to more accurately describe the process of multiple debonding processes occurring within a bulk particulate composite sample. This technique also provides an additional method to characterize adhesion strength, which allows for specific interrogation of surface modifications or binder additives targeted to increase adhesion. It is also suggested that this technique can be extended for use with non-transparent binders through the use of X-ray interrogation.

11.4 Conclusion

We have presented a methodology to characterize the debonding behavior between a binder and a material of interest. This was accomplished by extending an experiment conducted by Gent and Park [5] by incorporating Digital Image Correlation to obtain an in situ planar strain field measurement near the debonding event. We have successfully demonstrated that the strain information of the binder material surrounding a single particle leading up to and through a debonding event can be measured as a function of applied load. The strain field measurements allow for the characterization of interface strength as well as provide data useful to parameterize and validate computational models. Through experimental repetition over a large number of samples, distributions of debonding strength can be obtained for a given binder and particle material. Additionally, it is suggested that the method can be extended to investigate non-transparent binder materials through the use of X-ray interrogation. This technique is well-suited to quantify the effect of subtle changes in the adhesion at the particle-binder interface due to surface modification or the addition of surfactants.

Acknowledgments DISTRIBUTION A. Approved for public release; distribution unlimited. 96TW-2020-0040. This work was supported by research grant 17RWCOR469 from the Air Force Office of Scientific Research; Program Officer Dr. Martin Schmidt.

References

1. ASTM D4541-17: Standard Test Method for Pull-Off Strength of Coatings Using Portable Adhesion Testers. ASTM International, West Conshohocken (2017)
2. ASTM D1876-08(2015)e1: Standard Test Method for Peel Resistance of Adhesives (T-Peel Test). ASTM International, West Conshohocken (2015)
3. ASTM D6900-10(2015): Standard Test Method for Wet Adhesion of Latex Paints to a Gloss Alkyd Enamel Substrate. ASTM International, West Conshohocken (2015)
4. Tan, H., Liu, C., Huang, Y., Geubelle, P.H.: The cohesive law for the particle/matrix interfaces in high explosives. J Mech Phys Solids. **53**(8), 1892–1917 (2005)
5. Gent, A., Park, B.: Failure processes in elastomers at or near a rigid spherical inclusion. J Mater Sci. **19**(6), 1947–1956 (1984)
6. Lauke, B.: Determination of adhesion strength between a coated particle and polymer matrix. Composites Science and Technology. **66**(16), 3153–3160 (2006)
7. Lauke, B.: Stress field calculation around a particle in elastic–plastic polymer matrix under multiaxial loading as basis for the determination of adhesion strength. Composite Interfaces. **23**(1), 1–14 (2016)
8. Chu, T., Ranson, W., Sutton, M.A.: Applications of digital-image-correlation techniques to experimental mechanics. Experimental mechanics. **25**(3), 232–244 (1985)

Chapter 12
Preliminary Characterization of a Plastic Piezoelectric Motor Stator Using High-Speed Digital Holographic Interferometry

Z. Zhao, P. A. Carvalho, H. Tang, K. Pooladvand, K. Y. Gandomi, C. J. Nycz, C. Furlong, and G. S. Fischer

Abstract Precise surgical procedures such as deep brain tumor ablation may benefit from intra-operative image guidance using magnetic resonance imaging (MRI). However, the MRI's strong magnetic fields and constrained space pose the need for robotic devices to assist the surgeon. Piezoelectric motors are often used to actuate these robots. The piezoelectric resonant motor (PRM) is a class of such motors that consist of a bonded piezoelectric ring stator and a frictionally coupled rotor. Steady-state excitation at certain frequencies leads to specific mode shapes on the stator with surface waves having both in-plane and out-of-plane displacement components that cause the coupled rotor to spin. High-speed digital holography (HDH) can be used to measure time variant displacements with nanometer and microsecond resolution. We present initial measurements acquired at 67 k frame per second (fps) of two custom-designed plastic stator components operating at drive frequencies of 6.788 kHz and 6.980 kHz, respectively. Circumferential motion of the traveling surface waves with out-of-plane peak-to-peak displacement of approximately 100 nm peak-to-peak amplitude and a settling time of 2.99 ms was observed. Results demonstrate that plastic stators may be a promising alternative to metallic stators for use in the MRI environment.

Keywords MRI compatible actuation · High-speed digital holography · Image guided surgery · Resonant piezoelectric motor

12.1 Introduction

Over six hundred thousand cancer deaths were projected to have occurred in 2019 in the USA [1]. Cancer treatments have steadily advanced in the past decades and today includes immunotherapy [2], chemotherapy, radiation, and surgical intervention. In the latter, being able to leave adequate margins is paramount to reduce recurrence. Cancerous tumor recession surgeries have better long-term success rate when appropriate margins are achieved [3]. Magnetic resonance image (MRI) intra-operatively guided procedures have been shown to achieve better margins when compared to traditional methods [4].

The strong magnetic field and sensitivity to electrical noise impedes the use of most conventional actuators inside an MRI. Therefore, MRI compatible robotic devices use alternative actuators such as pneumatic [5, 6], hydraulic [7], or piezoelectric motors [8]. In the latter, non-ferromagnetic metals still remain. These metals can, in some cases, lead to distortions of the magnetic field and undesirable vibration due to gradient switching affecting image quality. Previous work has demonstrated that it is possible to substitute several of the components for plastic equivalents reducing magnetic field distortions without negatively affecting image signal-to-noise ratio [9]. In the aforementioned publication, the enclosure of the motor is modified

Z. Zhao · P. A. Carvalho (✉) · K. Y. Gandomi · C. J. Nycz
Robotics Engineering Program, Worcester Polytechnic Institute, Worcester, MA, USA
e-mail: zzhao4@wpi.edu; pacarvalho@wpi.edu; kygandomi@wpi.edu; cjnyz@wpi.edu

H. Tang · K. Pooladvand · C. Furlong
Mechanical Engineering Department, Worcester Polytechnic Institute, Worcester, MA, USA
e-mail: htang3@wpi.edu; kpooladvand@wpi.edu; cfurlong@wpi.edu

G. S. Fischer
Robotics Engineering Program, Worcester Polytechnic Institute, Worcester, MA, USA

Mechanical Engineering Department, Worcester Polytechnic Institute, Worcester, MA, USA
e-mail: gfischer@wpi.edu

M.-T. Lin et al. (eds.), *Advancement of Optical Methods & Digital Image Correlation in Experimental Mechanics*,
Conference Proceedings of the Society for Experimental Mechanics Series, https://doi.org/10.1007/978-3-030-59773-3_12

to reduce metal content; however, the copper stator remains the same. By replacing the stator with a plastic equivalent, the PRM should have a higher degree of MRI compatibility.

We extend our previous work in [10] by using a high-speed camera to view the start-up and initial settling behavior of a plastic piezoelectric motor stator component. The results show rotatory motion in the second mode of operation for the stator and demonstrate its potential in the construction of an MRI compatible PRM.

12.2 Methodology

The dimensional drawing of the plastic stator is shown in Fig. 12.1a. The plastic stator outside diameter is 30 mm and its height is 5.53 mm. A PZT-5H 30 mm (outside diameter) × 20 mm (inside diameter) × 0.5 mm (height) copper-plated ring (EBL Products, USA) was chemically etched to form the electrode pattern shown in Fig. 12.1b previously validated in [10]. To create the electrode geometry, a mask of the electrode pattern was printed and toner transferred onto the ring. Green toner reactive foil was used as the second mask to increase resolution of the transformed pattern as shown in [11, 12]. The pattern was subsequently etched using a Ferric Chloride solution. Figure 12.1c shows the etched ring bonded to a custom plastic stator manufactured from Ultem 1000 thermoplastic polyetherimide (PEI) using Loctite 3888 conductive epoxy. Four 90°-shifted sine waves were used to excite the crystal. The stator assembly with bonded piezo ring was secured onto a fixture designed to be mounted to an optical table.

By using a variation of the high-speed digital holographic (HDH) system presented in [13, 14], out-of-plane displacement and rotatory motion at the surface of the plastic stator can be observed. Figure 12.2 shows the HDH system setup for stator imaging using a 532 nm 30 mW laser (BWN-532-20E, B&W TEK INC, USA). The laser beam was split into beam *A* and beam *B*. Beam *A* was reflected by a mirror fixed to a piezo microactuator powered by a piezo controller (MDT694A, Thorlabs, USA). The reflected beam was focused into a fiber through a 20/0.40 lens. Beam B was reflected by a mirror and passed through a 60/0.85 lens to create a wavefront that was shown on the piezo stator. The resultant scattered wavefront was picked up by a 150 mm telecentric lens (Computar, USA) and reflected on a wedge. Beam A left the optical fiber and was shown on the other side of the same wedge. The combined beam was observed by a high-speed camera (FASTCAM SA-Z, Photron, USA) operating at 67 k frame per second (fps). An automated phase correlation calibration algorithm, applied in post-processing, that specifies the sampled image frames with the desired shift in interference phase during each high-speed phase ramp was designed. The system synchronized the PRM excitation phase and strobe phase by triggering automatically to reduce the frequency mismatch problem. To collect more holographic images in each cycle and better image view in each frame, low excitation frequency (the second mode) and voltage were operated.

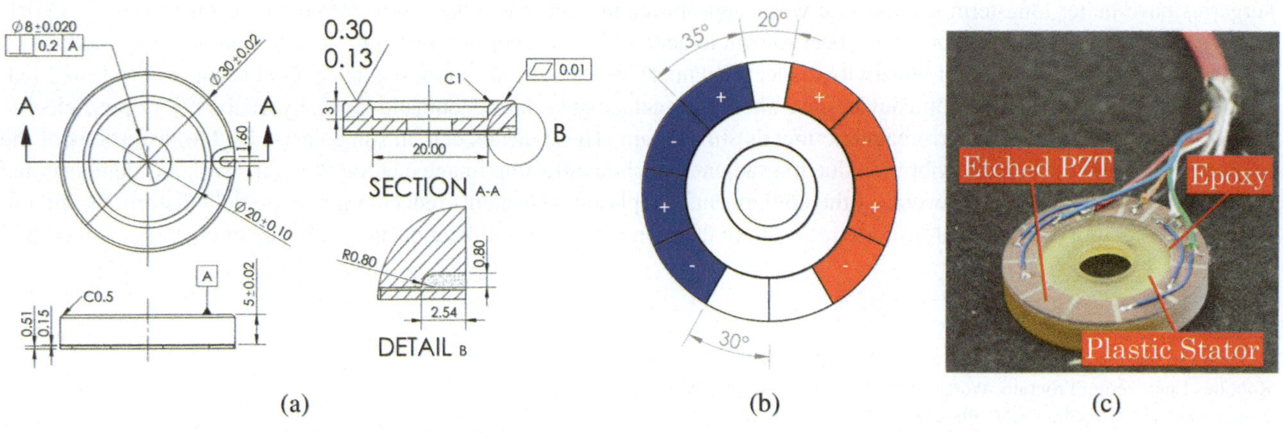

Fig. 12.1 (**a**) Dimensional drawings of the custom-designed and fabricated plastic stator. Outside diameter is 30 mm, and overall height is 5.53 mm. (**b**) Orthogonal view of underside of stator with wires coming out. (**c**) Electrode patterning of piezoelectric. Blue is sine, and red is cosine

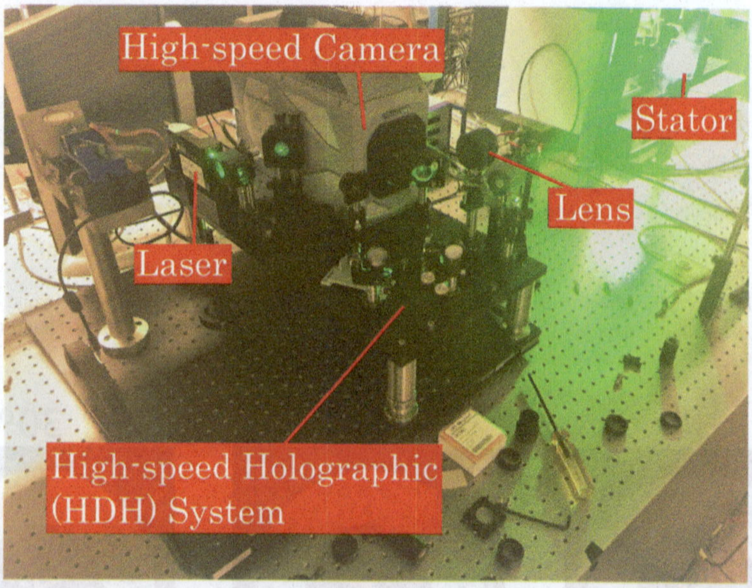

Fig. 12.2 Plastic stator and high-speed digital holographic (HDH) system on an optical table setup for quantitative imaging with nanometer resolution in full field to observe motor motion from startup to stability

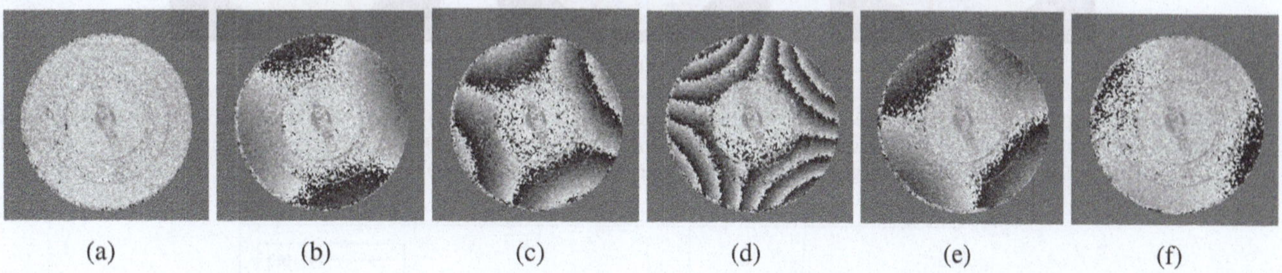

Fig. 12.3 High-speed digital holographic imaging of a plastic stator operating at 6.788 kHz and 240 V under normal operation mode. From **(a)**~**(f)**, rotatory motion along the stator surface can be observed. The time interval between consecutive images is one frame, which is approximately 14.9 μs

12.3 Experimental Results

We tested two equivalent manufactured custom plastic stators using the high-speed digital holographic (HDH) system. The two stators were operated in the second mode resonant frequency, respectively, at 6.788 kHz and 6.980 kHz, which yields approximately 10 frames per stator excitation cycle. Figure 12.3a–f shows 6 sequential frames from one of the plastic stators during a single excitation cycle operating at 6.788 kHz frequency with its surface waves moving circumferentially along the stator surface. The time interval between each of the two images is one frame, which is equal to 14.9 μs (1/67,000 s). The in-phase operation mode, which indicates the same sine waves-driven crystal setup, was tested and served as an experimental control. Figure 12.4a–f shows another equivalent manufactured plastic stator operating in an in-phase mode, since no rotating motion is expected or observed. The time interval is the same as the normal operation mode, which is 14.9 μs as well.

Raw results from holographic images are shown in Fig. 12.5. The stator showed rotatory motion during the experiment. From Fig. 12.5a and b (peak out-of-plane displacement), the wave peak rotated approximately 15° solid line position in estimate versus baseline. The peak moved approximately another 15° reaching a total rotation of 30° as shown by solid line in Fig. 12.5c. Notice that the stator operating frequency was 6.788 kHz, which is not perfectly matched to the high-speed camera operating frequency but is approximately 10 camera frames per stator excitation cycle. Under these circumstances, there will be a few frames mismatch and that will cause the resulting images to appear discontinuous.

The time-evolving out-of-plane displacement data was calculated from holographic images using the signal phase unwrapping methodology, and the result is shown in Fig. 12.6. The x-axis represents the frame number in the sequence, and one frame in the figure equals to 14.9 μs. The y-axis represents the Z component out-of-plane displacement in nanometers, with baseline at 50 nm (no relative motion) and 100 nm peak-to-peak amplitude. The Z field displacement versus time (frame)

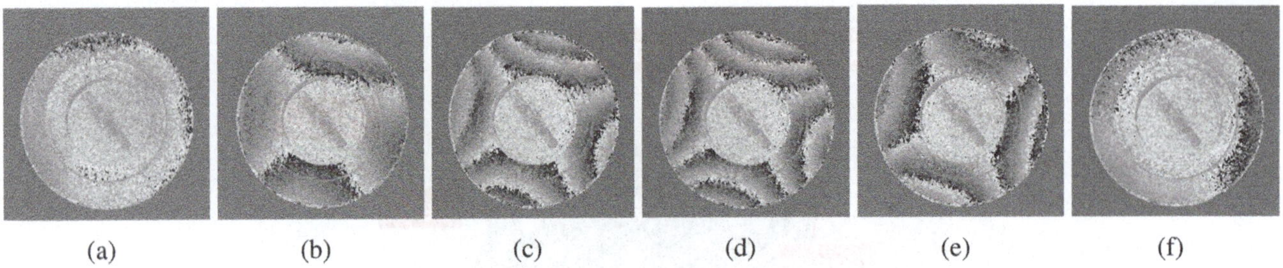

Fig. 12.4 High-speed digital holographic imaging of another equivalent plastic stator operating at 6.890 kHz and 160 V in an in-phase operation mode. From (**a**)~(**f**), it is shown that the stator has out-of-plane displacement without presenting rotation. The time interval between consecutive images is one frame, which is approximately 14.9 μs

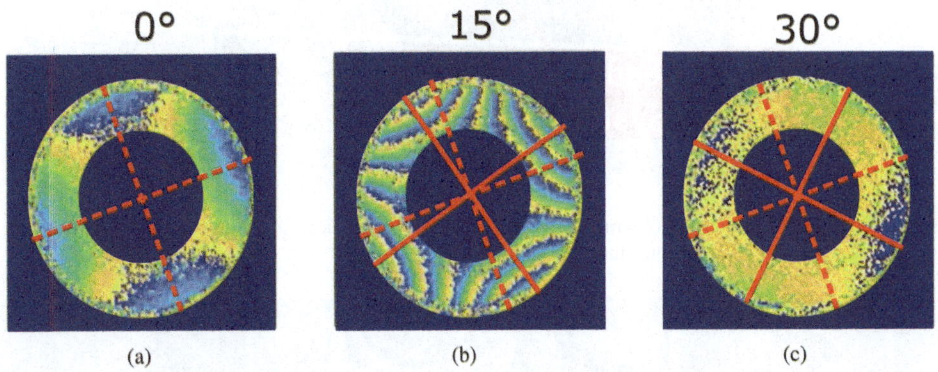

Fig. 12.5 (**a**) First appearance of out-of-plane displacement during steady-state excitation. (**b**) Peak out-of-plane displacement showing wave crest rotatory motion from initial position 0° dashed line to 15° solid line (approximately). (**c**) Out-of-plane displacement returns to baseline with rotatory motion (indicated by solid line) moving a total of 30° from initial position (dashed line)

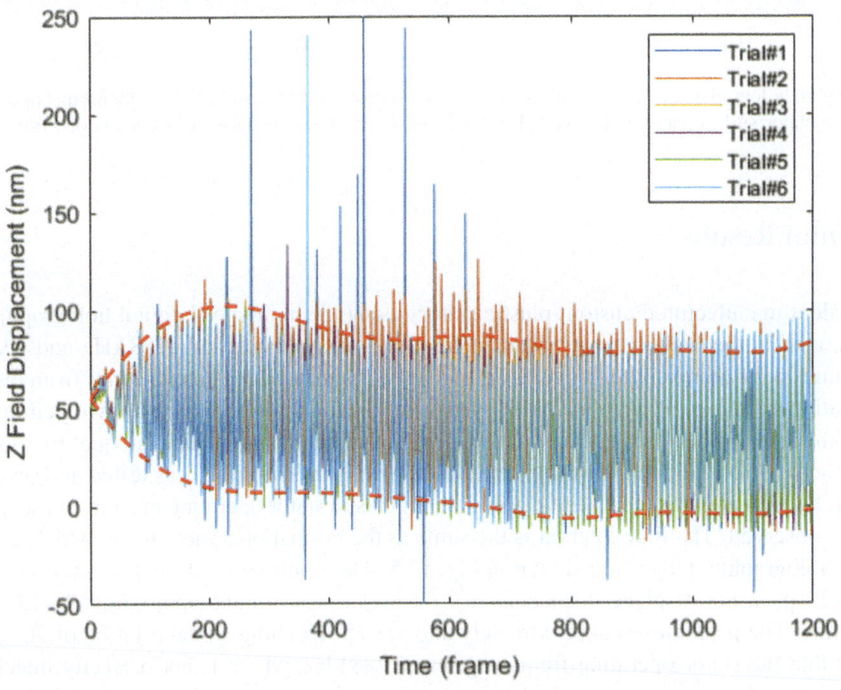

Fig. 12.6 Out-of-plane displacement over time from six trials with one plastic stator operating at 6.788 kHz and 160 V. Time series is similar for the different trials. Dashed lines represent the outer envelope. The stator vibrational amplitude envelope reaches peak at, approximately, the 200th frame indicating that the motor has settled into steady state

data from six trials with one plastic stator operating at 6.788 kHz and 160 V are shown in Fig. 12.6, and the overall shapes are similar to a Pearson correlation coefficient ranging from 0.8122 to 0.9615. The upper and lower red dashed lines indicate the envelope from effective data with outlier displacements removed. The stator is considered settled when the sine amplitude stops increasing. As observed, the stator settled to a steady state at, approximately, the 200th frame, which implies an estimated settling time of the amplitude of the envelop of the out-of-plane motion of 2.986 ms. Notice that since this work was operated at low excitation frequency and voltage, the displacement is much lower than expectation. By setting up higher excitation voltage to the crystal electrodes, the displacement will be larger compared to the data presented above.

12.4 Conclusion and Future Work

MRI compatible surgical robots are a promising avenue in the treatment of deep brain tumors. Motors minimizing the amount of conductive material may be needed to improve their compatibility with the scanner. An initial prototype of a plastic stator for a rotary piezoelectric actuator was imaged using a high-speed digital holography (HDH) system and compared to the metal stator of a commercially available motor. The ability of the stator to excite and sustain a stable mode points toward the possibility of using a plastic stator in the construction of a more MRI conditional piezoelectric motor.

However, some considerations are still required and are suggested for future work. Plastic is a poor conductor of heat, which may lead to overheating and damage of the bonded piezoelectric ring. The heating of the stator during operation should be quantified and other nonmetal materials, such as Macor glass ceramic, should also be evaluated. We have validated operation in the second mode. More commonly, piezoelectric motors operate in the fifth or sixth mode. The acquisition rate of the camera should be increased to enable measuring the motor with these higher frequency and voltage modes, which usually operate at frequencies from 30 kHz to 60 kHz requiring camera acquisition rates on the order of 600 k fps or higher. Determining a relationship between stator geometry and settling time is also left unexplored. Understanding these points may be beneficial prior to constructing a full motor assembly.

Acknowledgment Funding provided by the National Institutes of Health (NIH) under the National Cancer Institute (NCI) (Grant No. R01CA166379).

References

1. Siegel, R.L., Miller, K.D., Jemal, A.: Cancer statistics, 2019. CA Cancer J. Clin. **69**, 7–34 (2019)
2. Couzin-Frankel, J.: Cancer immunotherapy **342**(6165), 1432–1433 (2013)
3. Lisa, J.: Positive margins: the challenge continues for breast surgeons. Ann. Surg. Oncol. **15**, 1271–1272 (2008)
4. Christian, S., Andrea, B., Kea, F., Hartmut, V., Thomas, G., Volker, S.: Intraoperative MRI guidance and extent of resection in glioma surgery: a randomised, controlled trial. Lancet Oncol. **12**, 997–1003 (2011)
5. Dan, S., Danny, S., Doru, P., et al.: "MRI Stealth" robot for prostate interventions. Minim. Invasive Ther. Allied Technol. **16**, 241–248 (2007)
6. Fischer, G.S., Iulian, I., Csaba, C., et al.: MRI-compatible pneumatic robot for transperineal prostate needle placement. IEEE/ASME Trans. Mechatron. **13**, 295–305 (2008)
7. Rebecca, K., Kevin, L., Rao, G., et al.: Towards a teleoperated needle driver robot with haptic feedback for RFA of breast tumors under continuous MRI. Med. Image Anal. **13**, 445–455 (2009)
8. Nycz, C.J., Radian, G., Paulo, C., et al.: Mechanical validation of an MRI compatible stereotactic neurosurgery robot in preparation for pre-clinical trials. In: 2017 IEEE/RSJ International Conference on Intelligent Robots and Systems (IROS), pp. 1677–1684. IEEE, Piscataway (2017)
9. Carvalho, P.A.W.G., Nycz, C.J., Gandomi, K.Y., Fischer, G.S.: Demonstration and experimental validation of plastic-encased resonant ultrasonic piezoelectric actuator for MRI-guided surgical robots. In: ASME 2018 IMECE: V003T04A025-V003T04A025. American Society of Mechanical Engineers, New York (2018)
10. Carvalho, P.A., Haimi, T., Payam, R., et al.: Study of MRI compatible piezoelectric motors by finite element modeling and high-speed digital holography. In: Advancements in Optical Methods & Digital Image Correlation in Experimental Mechanics, vol. 3, pp. 105–112. Springer, Berlin (2020)
11. Mohammad, D., Donald, R., III, Zhanyue, Z., Matthew, S.: Increasing adhesion via a new electrode design and improved manufacturing in electrostatic/microstructured adhesives. J. Electro. **91**, 48–55 (2018)
12. Dadkhah, M., Zhao, Z., Wettels, N., Spenko, M.: A self-aligning gripper using an electrostatic/gecko-like adhesive. In: 2016 IEEE/RSJ International Conference on Intelligent Robots and Systems (IROS), pp. 1006–1011. IEEE (2016, October)
13. Haimi, T., Payam, R., Koohyar, P., et al.: High-speed holographic shape and full-field displacement measurements of the tympanic membrane in normal and experimentally simulated pathological ears. Appl. Sci. **9**, 2809 (2019)
14. Payam, R., Haimi, T., Rosowski, J.J., Cosme, F., Cheng, J.T.: Combined high-speed holographic shape and full-field displacement measurements of tympanic membrane. J. Biomed. Opt. **24**, 031008 (2018)

Chapter 13
DIC Measurement of Anisotropy for Plastically Deformed Thermoplastic

Kenichi Sakaue and Sho Higuchi

Abstract The present research examines the deformation-induced anisotropy in the necking part of thermoplastic. Firstly, a uniaxial tensile force is applied to the large polycarbonate specimen to create the necking part. Then, small specimens are cut off from the large deformed specimen with various cutting angle to the initial tensile direction. The small specimens are used to evaluate the stress-strain curves. The results show that the elastic modulus and the maximum stress depend on the cutting angle to the initial tensile direction. Therefore, it is confirmed that the plastic deformation-induced anisotropy. Moreover, it is revealed that the dependence of the cutting angle on the elastic modulus and the maximum stress can be expressed by the anisotropic elastic constitutive equation and Tsai-Hill criterion, respectively.

Keywords Thermoplastic · Necking · Deformation-induced anisotropy · Elasto-plasticity · Digital image correlation

13.1 Introduction

The necking phenomenon of thermoplastic is frequently observed if the thermoplastics specimen is subjected to a uniaxial tensile force. It is well-known that the deformation-induced anisotropic characteristics appears in the necking part by the orientation of the molecular chains. The deformation-induced anisotropic characteristics of the thermoplastics have been studied through many experiments. These researches revealed that the elastic modulus and the maximum stress increases in the extending direction, and these characteristics depends on the value of the pre-strain [1]. Also, the researches on the numerical analyses have been performed to reproduce the deformation-induced anisotropy in thermoplastics materials. For example, the molecular chain network theory [2] and the composite inclusion model [3] are proposed for the finite element simulations.

As mentioned above, the deformation-induced anisotropy of thermoplastics has been investigated by many studies mainly through experiments. However, the purposes of the most researches are to examine the existence of anisotropy and to evaluate the relationship between pre-strain and anisotropy. Therefore, the strain distribution has not been measured. It is necessary to acquire the strain distribution by full-field strain measurement methods in order to confirm the validity of the anisotropic constitutive model of the deformed materials and the method of finite element analysis. In this study, a uniaxial tensile test is performed to the specimen cut off from the necking part of the large polycarbonate sheet. In order to obtain the strain distribution, digital image correlation method, DIC, is used. In addition, the anisotropic mechanical characteristics are evaluated by the anisotropic elastic constitutive equation and Tsai-Hill criterion, respectively.

13.2 Test Method

Polycarbonate sheet with 3 mm thickness (Iupilon, Mitsubishi Engineering Plastics Co., Ltd) is used as the test material. Firstly, the large specimens are used to create deformation-induced anisotropic characteristics in the necking part by a uniaxial tensile test. Then, the small specimens are cut off from the necking part of the large specimen and are used to

K. Sakaue (✉) · S. Higuchi
Department of Mechanical Engineering, Shibaura Institute of Technology, Koto-ku, Tokyo, Japan
e-mail: sakaue@sic.shibaura-it.ac.jp; md19059@shibaura-it.ac.jp

M.-T. Lin et al. (eds.), *Advancement of Optical Methods & Digital Image Correlation in Experimental Mechanics*, Conference Proceedings of the Society for Experimental Mechanics Series, https://doi.org/10.1007/978-3-030-59773-3_13

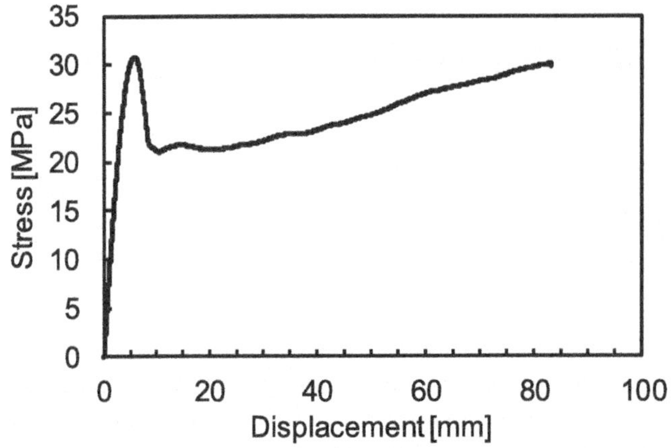

Fig. 13.1 Stress-displacement curve in tensile test for a large polycarbonate specimen

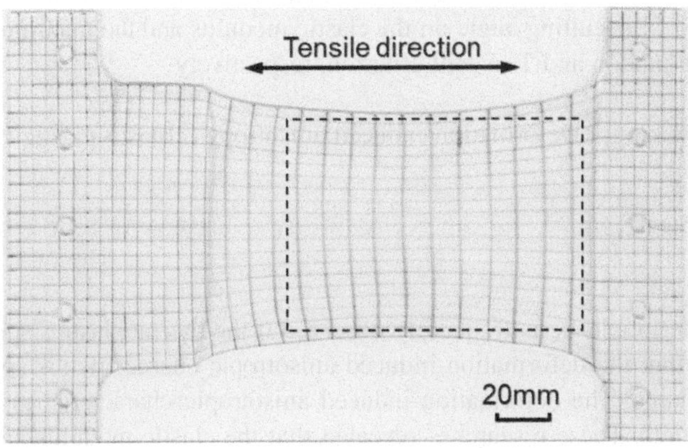

Fig. 13.2 Image of the large specimen after tensile test

evaluate the plastic deformation-induced anisotropy. The size of the large specimen is about 190 mm length and 150 mm width, and the size of the small specimen is about 45 mm length and 12 mm width.

A uniaxial tensile test for the large specimen is performed by a tensile testing machine with a thermostatic-chamber. In order to reduce the tensile force to create the necking part in the large specimen, the tensile test is carried out at a temperature of 120 °C and at a crosshead displacement speed of 2 mm/min. Figure 13.1 shows the typical nominal stress-displacement curves of the large specimen. The stress reaches the maximum value of about 32 MPa subsequently rapidly and decreases to about 23 MPa. Then, the stress increases gradually with the propagation of the necking part. Figure 13.2 is the image of the large specimen after the uniaxial tensile test. A 5 mm × 5 mm grid pattern was drawn before the tensile test. It is observed large plastic deformation by the propagation of the necking part. The small tensile specimens are cut off from the necking part in the large specimens by using an NC milling machine. The small specimens are cut from the area enclosed by dashed line in Fig. 13.2. The cutting direction from the necking part changes by 15-degree increments from the tensile direction (0 degree) of the large specimen to the perpendicular direction to the tensile direction (90 degrees).

The uniaxial tensile test for the small specimen is carried out in a thermostatic chamber. The temperature condition is 25 °C, and the crosshead displacement speed is 1 mm/min. The displacement and strain on the specimen surface is measured by digital image correlation, DIC. The images are taken with a monochromatic CCD camera at an interval of 1 s.

13.3 Results

In the present study, the nominal stress and the nominal strain are used to evaluate for the stress-strain curve. Figure 13.3 shows the stress-strain curves evaluated from the small specimens cut off in various directions from the necking part in the large specimens. The stress-strain curves of the cutting angles of 0, 15, and 30 degrees show the typical stress-strain curve of

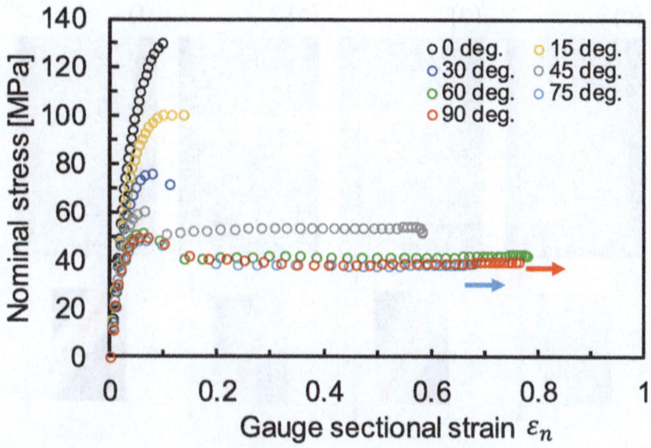

Fig. 13.3 Stress-strain curve evaluated from small specimens cut off from the necking part

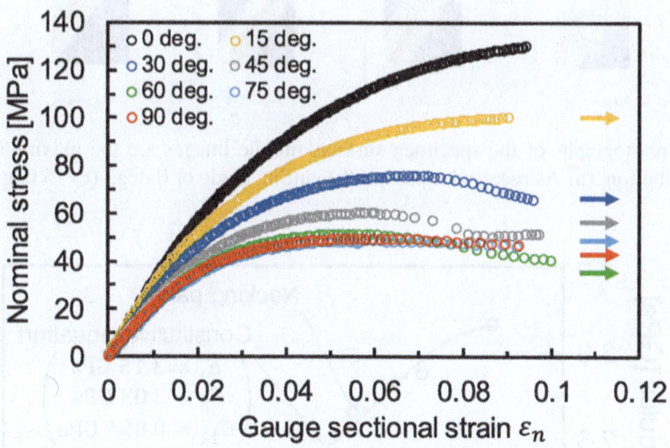

Fig. 13.4 Initial part of the stress-strain curve in Fig. 13.3

brittle materials. On the other hand, the stress-strain curves of the cutting angles of 45, 60, 75, and 90 degrees represent that the stress reaches each maximum value and then stress decreases to each constant stress value and strain continue to increase at the constant stress value. The necking is also observed in these specimens. The specimens with the cutting angles of 45 and 60 degrees fractured finally, but the specimens of 75 and 90 degrees did not fracture at the last strain measurement as arrows is indicated. As shown in Fig. 13.3, the maximum stress decreases with increasing the cutting angle from the necking part in the large specimens.

Figure 13.4 shows the enlarged figure of the initial part of the stress-strain curves in Fig. 13.3. Figure 13.4 indicates that the initial stiffness depends on the cutting angle from the necking part. The elastic modulus is the largest in the direction of 0 degree. And the elastic modulus decreases with increasing the cutting angle from the necking part in the large specimens. In the present study, it is confirmed that the material as received is isotropic mechanical characteristics by the same uniaxial tensile tests. Therefore, the anisotropic mechanical characteristics appear in the necking part by plastic deformation.

Figure 13.5 shows the images of the small specimens, the maximum strain, and the minimum strain distributions at the nominal strain of about 0.1. The maximum and the minimum strain in the small specimen of 0 degree direction is approximately uniform distribution. As shown in Fig. 13.5, the strain distribution in the small specimens with the cutting angle of 15, 30, 45, 60, and 75 degrees are not uniform, and the large strain band appears obliquely from bottom right to the upper left. The anisotropic elastic analysis represents that the large strain band is appeared obliquely in the specimen if the anisotropic elastic materials is subjected to uniaxial tensile force in the different direction to the principal axes of anisotropic elastic materials. Therefore, this experimental results correspond with the result of anisotropic elastic analysis.

Figure 13.6 shows the relationship between the elastic modulus and the cutting angle from the necking part in the large specimens. The elastic modulus is calculated from the slope of the stress-strain curve in the range from 0.001 to 0.01. The results evaluated from the material as received are also shown in Fig. 13.6. It is found that the material as received is isotropic. On the other hand, the elastic modulus in the necking part decreases with increasing the cutting angle. Therefore, the

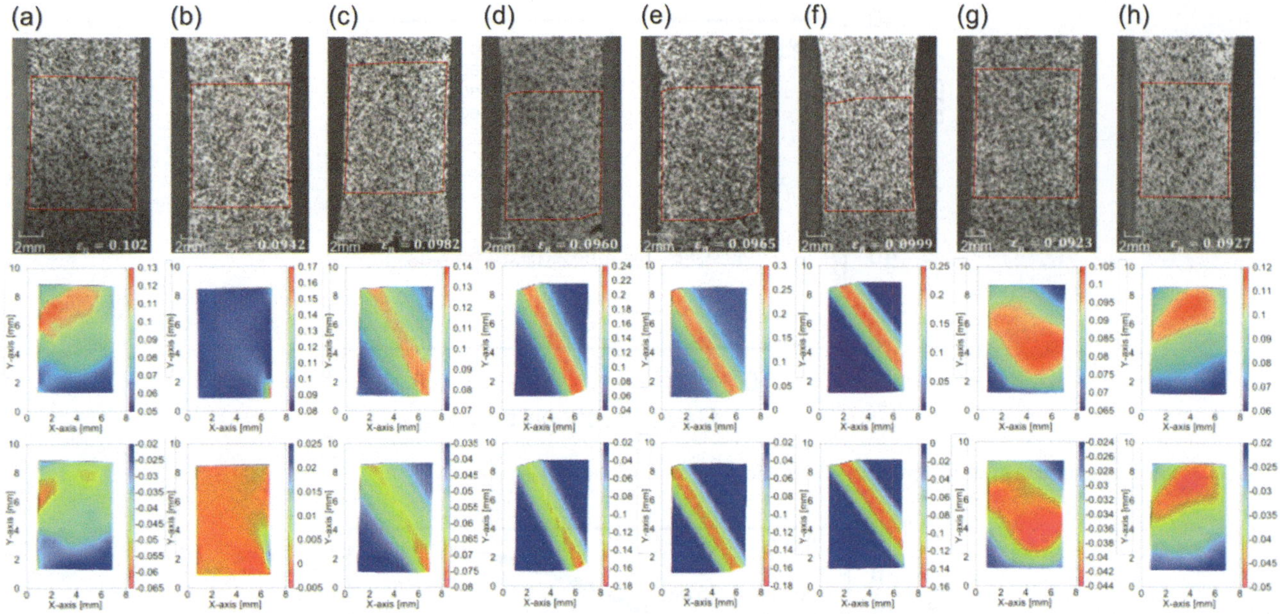

Fig. 13.5 Upper images are of the photographs of the specimen surface, middle images are the maximum strain distribution, and the bottom images are the minimum strain distribution. (**a**) As received material, (**b**) cutting angle of 0 deg., (**c**) 15 deg., (d) 30 deg., (**e**) 45 deg., (**f**) 60 deg., (**g**) 75 deg., and (h) 90 deg.

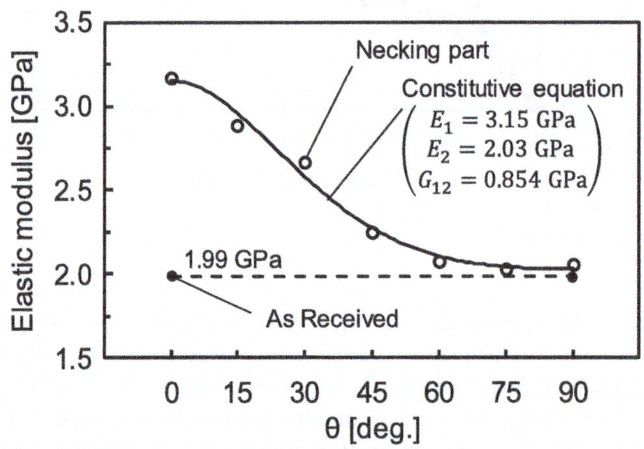

Fig. 13.6 Relationship between elastic modulus and cutting angle

deformation-induced anisotropy appears in the necking part by plastic deformation. The elastic modulus E_1' in the direction of angle θ is given by coordinate transformation of the anisotropic elastic constitutive equation and is expressed by

$$\frac{1}{E_1'} = \frac{\cos^4\theta}{E_1} - \frac{2\nu_{12}\cos^2\theta\sin^2\theta}{E_1} + \frac{\sin^4\theta}{E_2} + \frac{\cos^2\theta\sin^2\theta}{G_{12}} \tag{13.1}$$

where E_1 and E_2 are the elastic modulus in the direction of 0 and 90 degrees, respectively, and G_{12} is shear modulus. Thus, E_1, E_2, and G_{12} are determined by Newton-Raphson method using the elastic modulus E_1' in arbitrary directions. Here, the value of Poisson's ratio ν_{12} is supposed to be 0.35. The solid line in Fig. 13.6 is calculated elastic modulus in arbitrary direction θ from determined E_1, E_2, and G_{12}. The anisotropic elastic constitutive equation is in good agreement with the elastic modulus evaluated by the experiments. Therefore, the anisotropic elastic constitutive equation can represent the elastic modulus in arbitrary direction caused by plastic deformation-induced anisotropy.

Figure 13.7 shows the relationship between the maximum stress in the stress-strain curve and the cutting angles. Figure 13.7 also indicates the maximum stress evaluated from the material as received. As shown in Fig. 13.7, the tensile strength in the material as received is isotropic. On the other hand, the maximum stress in the necking part decreases with increasing

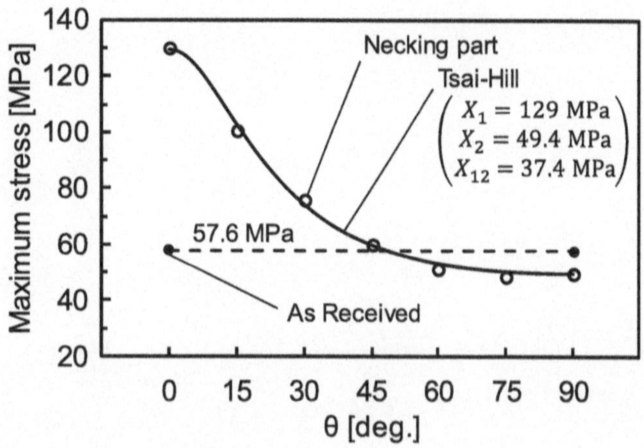

Fig. 13.7 Relationship between the maximum stress and cutting angle

the cutting angle. Therefore, the deformation-induced anisotropy of the maximum stress appears in the necking part by plastic deformation. In order to express the anisotropic plasticity, the stress σ in the direction of angle θ is given by Tsai-Hill criterion and is expressed by

$$\left(\frac{\cos^4\theta}{X_1^2} - \frac{\cos^2\theta\sin^2\theta}{X_1^2} + \frac{\sin^2\theta}{X_2^2} + \frac{\cos^2\theta\sin^2\theta}{X_{12}^2}\right)\sigma^2 = 1 \tag{13.2}$$

where X_1 and X_2 are the tensile strength in the direction of 0 and 90 degrees, respectively, and X_{12} is shear strength. Thus, X_1, X_2 and X_{12} are determined by Newton-Raphson method using the maximum stress σ in arbitrary directions. The solid line in Fig. 13.7 is the stress in arbitrary direction of angle θ expressed by Tsai-Hill criterion calculated from determined X_1, X_2, and X_{12}. Tsai-Hill criterion is in good agreement with the maximum stress evaluated by the experiments. Therefore, Thai-Hill criterion can represent the maximum stress in arbitrary direction caused by plastic deformation-induced anisotropy.

13.4 Conclusion

A uniaxial tensile test is carried out for the specimen cut off from the necking part of the large polycarbonate sheet. The test results show that the elastic modulus and the maximum stress decrease with increasing the cutting angle from the necking part. The anisotropy of the elastic modulus and the maximum stress can be expressed by the anisotropic elastic constitutive equation and Tsai-Hill criterion, respectively.

References

1. Senden, D.J.A., Van Dommelen, J.A.W., Govaert, L.E.: Strain hardening and its relation to Bauschinger effects in oriented polymers. J. Polym. Sci. B Polym. Phys. **48**, 1483–1494 (2010)
2. Tomita, Y.: Constitutive modelling of deformation behavior of glassy polymers and applications. Int. J. Mech. Sci. **42**, 1455–1469 (2000)
3. Mirkhalaf, M., Van Dommelen, J.A.W., Govaert, J.E., Furmanski, J., Geers, M.G.D.: Micromechanical modeling of anisotropic behavior of oriented semicrystalline polymers. J. Polym. Sci. B Polym. Phys. **57**, 378–391 (2019)